How to Make Great Music Mashups

Written for the beginner DJ, this accessible book presents everything you need to know in order to create great dance floor moments that will take your sets to the next level and get you noticed as a DJ. Using Ableton's industry-leading digital audio workstation, the reader will learn to achieve a professional sound by expertly manipulating Warping, pitching, editing, automation and plugin effects processing; also, avoiding mistakes such as key-clashing, jarring transitions, mismatched energies and more. The book's companion website includes key-charts, musical scale diagrams, organisational templates for live sessions, and follow-along video demonstrations.

Paul Zala has over 25 years of musical experience behind him, covering everything from training in piano, guitar, trumpet and voice through to performance in live bands and DJing. Coupled with training in Music Production at RMIT Melbourne, he's been writing, producing, engineering and performing professionally since 2007. Finding his feet as one half of Australian Dance outfit Denzal Park, he's enjoyed a successful dance music career sporting dozens of singles, around 100 remixes, international performances, not to mention accolades including APRA Dance Work of the Year, Platinum Records, number 1's on the US Billboard Dance Chart, Australia's ARIA Club Chart and the top spot on dance music's leading outlet Beatport.com. Paul loves talking shop and sharing his knowledge with anyone who loves making music.

How to Make Great Music Mashups

The Start-to-Finish Guide to Making Mashups with Ableton Live

Paul Zala

Routledge
Taylor & Francis Group

NEW YORK AND LONDON

First published 2018
by Routledge
711 Third Avenue, New York, NY 10017

and by Routledge
2 Park Square, Milton Park, Abingdon, Oxon, OX14 4RN

Routledge is an imprint of the Taylor & Francis Group, an informa business

© 2018 Taylor & Francis

The right of Paul Zala to be identified as author of this work has been asserted by him in accordance with sections 77 and 78 of the Copyright, Designs and Patents Act 1988.

Library of Congress Cataloging-in-Publication Data
A catalog record for this book has been requested

ISBN: 978-1-138-09276-1 (hbk)
ISBN: 978-1-138-09278-5 (pbk)
ISBN: 978-1-315-10728-8 (ebk)

Typeset in Giovanni Book
by Florence Production Limited, Stoodleigh, Devon, UK

Visit the website: www.makegreatmusicmashups.com

Contents

About the Author

Australian-born Paul Zala has been learning, playing and writing music almost his whole life. Having studied piano, guitar, trumpet and singing, as well as musical theory and music technology, his discovery of dance music over a decade ago has seen him turn his more classical training toward the pursuit of modern club music. Since then he has written, produced, remixed and performed as a producer and DJ, spending most of his dance music career as one half of Australian duo Denzal Park. During his time in the duo he has achieved number one spots on Beatport, club charts and Billboard Dance charts with Denzal Park's original tracks, platinum selling singles produced for commercial artists, as well as writing over a hundred remixes, including mixes for dance cross-overs such as Kylie Minogue, Avicii and Dirty South.

Paul has performed DJ sets across Australia and over the world, and continues to work in music every single day.

Acknowledgements

I'd like to thank the DJs who kindly contributed their well-founded opinions and experience on the subject of mashups: Kam Denny, Ivan Gough, James Ash, Phil Ross, Pierce Fulton, Anthony Trimboli, Dave Winnel and John Walden. I respect you all as DJs and producers, and your comments opened up new ideas to pursue in the book.

Screenshots of Ableton Live appear courtesy of Ableton AG who generously allowed me to write about their industry-leading software: https://ableton.com.

Thanks to Mixed In Key, who granted permission to mention their unique software in the book and for the use of the Mixed In Key Camelot wheel. Acknowledgements also to Xfer Records and vladg/sound for permission to show screenshots of their respective audio plugins.

Thanks especially to my long-time production partner Kam Denny who taught me a lot about producing and DJing in general, particularly some of the more crucial concepts behind making mashups that can actually work on the dance floor. Acknowledgements to my brother Tim for running his eyes over sections of the book, and my wife Bree for supporting me writing it and just being who she is.

Finally, this book is dedicated to all the hard-working and talented musicians without whom there could be no such thing as mashups.

Introduction

Let me start with some questions. Do you want to know how to create better, more powerful, more effective mashups? Do you want to give your audience more creative, memorable moments? Do you want to stand out from other DJs, become more versatile, and maximise your potential to get more gigs? If you're reading this, then the answer is probably *yes*.

Mashups are a great tool not only for enhancing your DJ sets, but also for teaching you about music and making you think about what makes an audience respond.

Since you are reading this, you're most likely in one of two groups. Either you're just starting out with mashups, you can see the great potential of them, but you just don't know how to make them as professional or effective as the ones that the big DJs play. Or you already create your own mashups—but you are hungry for more techniques, shortcuts and tools that will take yours to the next level.

Fortunately, this book is designed to help both. Whether you are just getting started or you are an experienced DJ, this book will help you get the best out of your mashups, and take them from simply being functional to being an entire art-form in themselves.

After reading this book you should better understand the purpose of mashups and the decisions behind what music you choose to pair together. You'll understand the roles of the different elements used in mashups, the musical and emotional connections between different songs, and why certain things work on the dance floor. We'll also go into specifically using the best software for the job—Ableton Live. We'll explore the software and techniques used, and how to keep things organised to give you the best chance of finding killer ideas and turning them into reality, *fast*.

So why listen to me? Good question. At the time of writing this book I've been writing and producing dance music for over ten years, as well as touring as a DJ around my home country of Australia most weekends, sometimes internationally. Mashups play a *huge* part for me as a DJ, and I love making them. I have been fortunate enough to count some of Australia's most talented dance music producers and mashup creators as my friends, many of whom have kindly shared their ideas and lessons, which I have included in the book.

I initially got into writing mashups so that I could give commercial club audiences a taste of great club music, while giving them elements of commercial tracks they felt safe with, so that it wouldn't scare them off the dance floor. If I could convince an inner city top-40 nightclub to stay on the dance floor while playing two hours of cutting-edge club music by making it feel *as one* with their favourite songs, that was a win for me. But there's so much more you can do. As time has gone on, dance music has become intertwined with mashup culture to the point where the dance floor now expects to be taken on a journey of collaged music.

I've spent a *lot* of my time creating mashups (some might say too much time) and seeing their reactions on a dance floor. After initially making my fair share of messy, badly executed ideas, I found that a lot of the important lessons I learned about mashups actually came from learning how to produce dance music. While out at gigs, I used to hear a lot of mashups by DJs who had brilliant ideas, but couldn't quite make them work. Coming from a musical background, it took me some time to realise that many of the musical concepts required to make mashups were difficult for DJs to learn or find information on. After studying mashups more I discovered that many didn't yet know how to account for concepts such as key-matching and energy-matching, or they experienced arrangement problems and mixing issues. They obviously knew what they wanted to hear, they just didn't know how to make it reality yet. On the other end of the scale, I've heard electronic music acts that transform mashups from a simple tool to appease commercial crowds to a new form of DJ set completely, weaving one piece over another and creating a mind-blowing musical patchwork.

After years of having DJs ask me about mashups, I decided it was time to put *everything* that I've learned over the years into this book. I want to give you the keys to help you turn your greatest ideas into even greater dance floor moments.

Anyway, thank you for your time and attention, and I hope this book can help and motivate you to *Make Great Music Mashups*.

CHAPTER 1

Why Mashups?

What is a mashup? Why would you use one? Is it difficult to put together?

Chances are if you're reading this book, you already have some experience with mashups, whether by making your own, or hearing some effective mashups in sets from your favourite DJs. Some DJs even base their whole careers off mashups! So let's begin by explaining . . .

WHAT IS A MASHUP?

Put simply, a mashup is the blending of two or more sections of music to create something new. This can be as simple as a **vocal** from one song placed over the instrumental version of another; or it can be as complex as a five-song mashup complete with a vocal from one song, the beats and **bass-line** from another song, an extra hook melody from another, with a vocal chant from yet another. Bear in mind, the latter idea may sound terrible and be too much for the listener to take in, but we'll get to that. The important concept here is that if a DJ has picked the right sections of music to blend, it will sound as if they were *made to exist together*, creating a whole new piece of music that is unique to you.

Mashups unofficially entered existence in the 1970s and 80s as musical performances or recordings where artists would blend melodies or vocals from one song with another. Rather than using recordings, they would actually be performed this way, usually only included as novelty performances or bonus tracks on albums.

It took on a more familiar form once record labels started releasing dance music records with an **acapella** version (vocal-only version) included on the vinyl. DJs would play the acapellas over other songs as part of their DJ set, similar to what we hear today in pre-recorded mashups. Soon, producers began using these acapellas in their productions to create unofficial **remixes**, or 'bootlegs'.

As computers and compact discs came along, DJs discovered that they could prepare their ideas and bring them along to gigs the very same night. By studying the arrangement of their elements and editing them to coincide at the right times, they could be much more creative and craft much better moments; almost new pieces of music. By working on dedicated mashup projects in the studio, they found they could get better results than simply playing things over the top of each other live and hoping for the best. They began playing them in their sets, and even distributing them online, often illegally.

As mashup culture started to take off in the United Kingdom and then spread to Europe and the United States, it gained a lot of exposure in 2004 when American DJ/Producer Danger Mouse released his experimental mashup album 'The Grey Album', **sampling** from only two sources: The Beatles' 'White Album' and vocals from Jay-Z's 'The Black Album'.

In more recent times, DJs usually employ the art of mashups to present an audience with the newest club music as well as their own music (if they have released any). However, some electronic music acts that just play their own songs incorporate the idea of 'mashing up' their music to create a unique take on their repertoire, and give the fans an experience they could not recreate at home.

These days, mashups have pretty much become standard to have in your DJ set. If you are not producing your own tracks, they can be your ticket to playing a DJ set that stands out.

> I remember when I was just starting out, hearing a DJ run the vocal of 'Dub Be Good To Me—Beats International' over the b-side dub of Wreckx-N-Effect 'Rump Shaker', and then dropping to just the a-cappella and bringing in another 3rd song, and I was like 'Wow . . . that's impressive!' Everything worked together perfectly as the keys complemented nicely and the DJ's performance was very tight.
>
> (Kam Denny, Veteran Australian DJ/Producer)

WHY WOULD YOU USE ONE?

There are many reasons to use mashups. Here are a few.

Standing Out from Other DJs

Having a collection of mashups on hand makes you more versatile, able to fill any time slot and hold a dance floor for an extended period of time. They also possess their own 'flavour', giving your set a sound that is unique to you.

> Sometimes DJ sets are only 45 minutes or an hour now. You've got so many DJs trying to make their mark; this is how you set yourself apart.
>
> (Phil Ross, Long-time commercial club DJ)

> Don't continue playing other people's mashups. Yours will help YOU stand out as a DJ.
>
> (James Ash, Veteran DJ/Producer)

Creating Moments

When a DJ blends two or more pieces of music together, they borrow something from each, but they also create something the original artists didn't intend. For instance, an aggressive, big-room electro track is building to a massive peak, when suddenly, yet quite musically, the track **transitions** into an incredibly emotional breakdown with a classic house vocal. If done well, the DJ has delivered the audience a big moment they did not expect.

Mashups also enable you to simply increase the number of moments you have in your set. Whilst this isn't always the goal, modern sets tend to be relentlessly jam-packed full of moments. But as the saying goes, with great power comes great responsibility. Just remember that a few high-quality moments are always more effective than squashing in as many moments as you can.

> I remember when I first heard of mashups by guys like Girltalk or Milkman and being so excited because it's basically like an ice cream sundae of good music.
>
> (Pierce Fulton, US DJ/Producer)

> Watching Madeon and the way his music shifted from song to song—I almost felt like every record was a mashup of at least two track or more, and the seamless transitions between them as well as the energy he created was incredible. That was definitely an eye-opening lesson for me in how creative one could be with mashups, whether it be pre-organised or done on the fly in a DJ set.
>
> (Walden, DJ and Producer Europe/Australia)

Giving People a Taste of the Familiar or Breathing New Life into Older Music

By playing a well-known or current vocal, the DJ is able to push the audience out of their comfort zone while also giving them something that they can relate to. They can play something alongside it that would ordinarily have scared club-goers off the dance floor, such as music that is too emotional, hard-hitting, or simply too underground. Similarly, the DJ can use mashups to bring great musical moments into a set they could never normally exist in. A prime example would be bringing in a section from a well-known rock track to act as a **breakdown** for a dance track with a similar energy, such as a pacey electro track. This combination has enabled the DJ to bring the feeling of a completely different genre of music into a dance set without feeling completely foreign.

> I like to use a mashup to connect a current record with a classic. It's a great way to have the crowd interact with a record that they know, and one that they don't.
>
> (Ivan Gough, Veteran DJ and Producer)

> Breathe more life into an old track, or make a brand new track more familiar to audiences who haven't heard it before.
>
> (James Ash, DJ/Producer)

Giving New Meaning to a Musical Element

With good use of mashups, you can transform the meaning of an element, putting it in a new context for the listener. The most effective use of this theory is when dealing with vocals, which communicate very little harmonic or rhythmic information on their own. Depending on what you put underneath it, a soft pop vocal can suddenly feel like a classic house vocal; a dreamy, happy vocal in a **major** key can be heard as a dark, brooding performance in a **minor** key. More on how this technically works later on in the book.

> Mashups are a great way of pulling different material together to bring a new energy and interpretation on those tracks. I like to think of it as akin to sampling old records or even just remixing in general.
>
> (Walden, DJ/Producer)

Creating a Feeling of International Production and Giving it Your Stamp

DJ/producers who bring mashups with their own tracks' vocals and melodies to their shows end up putting on a show that feels distinctly 'theirs'. This is not only because you hear parts of their songs during the set, but also because the way they choose to blend them will naturally feel similar to the way they choose to put together and mix their own tracks. Even if you aren't a producer, adding mashups to your skillset is an effective way to help to solidify 'your sound' as a DJ.

> Create a 'sound' that is unique to you, even if you're playing popular music.
>
> (Phil Ross, Commercial DJ)

At first, mashups may seem like a confusing concept. There are many things to consider, and if the sections don't blend well, it can result in some pretty monstrous sounding music. Follow the simple steps in this book, and you'll be well on your way to avoiding many of the easy-to-make mistakes that come with creating mashups. This book will teach you which parts of the Digital Audio Workstation 'Ableton Live' you need to be familiar with, how to treat your audio, which techniques and effects you should start with to make sure your mashups make sense, and how you can avoid mismatches and arrangement mistakes that confuse the listeners and lose the dance floor.

IS IT DIFFICULT TO PUT A MASHUP TOGETHER?

Most DJs who initially have difficulties with mashups struggle with picking the right elements to go together. When it comes to rhythm, they usually have some kind of instinctive feel, which is why they are DJs in the first place, and what enables them to beat-mix and play music in an order that makes sense. **Key**, however, is much harder to wrap your head around, particularly if, like many DJs, you come into DJing because of a love of the scene rather than because you are a producer or musician. Once you understand key, you'll be in a much better place.

The rest is just practice. Once you have the techniques down and have the muscle memory to whip around Ableton Live without thinking, you'll be able to craft mashups in no time. In the beginning, it seems like a very confusing process, but most of the techniques I talk about in this book are just little tricks, and although there are many, most of them are working to achieve the same thing; a sense of unity that helps keep the listener engaged with the music.

So, read on. Once you have created a few, I promise it will become very quick and easy.

AN IMPORTANT CONCEPT ABOUT THE DANCE MUSIC AUDIENCE

Before we go on, there is an important concept I want you to start thinking about that affects almost every technique in this book. You'll notice that I keep coming back to it. It's a concept that my long-time production partner Kam Denny really drilled into me when we first started producing music together. It's all about giving elements *space* and directing the listener's attention.

When you are in the middle of producing, whether it is a full track or a mashup, you tend to get lost in the creation of it and forget to 'hear the music' as a person on the dance floor will hear it. Often, they are hearing a lot of the music for the first time, particularly during a DJ set where playing brand new tracks is important.

The human ear (or really, the brain) can only pay attention to a certain amount of musical elements happening at once. It can also only pay attention to one or two new sounds starting at the same time. After that, the mind needs a moment to 'settle into' these elements, to get a feel for how they sound and what kind of pattern they make. Even if you have an amazing combination of elements to craft into a mashup, putting too many elements together at once will only create confusion for the listener, and confusion is the fastest way to empty your dance floor.

If you have a very important element that requires a lot of listener attention (such as a vocal), it needs to be given the space it deserves so that you can be 100 percent sure that everyone's ears will be focused on it when it arrives. It is important to also remember that any fully produced track you use may only look like one **clip** on your screen, but it is already made up of many sounds, each one already fighting for the listener's ear.

As we go through this book, you will see that almost every technique is designed to help guide the listener through the music without confusion and clumsiness. This is what great mashups all have in common. You may also notice that many of the mistakes new mashup artists make relate to this concept of space.

Guiding the listener with pace and control is especially important for club music because the musical journey in a nightclub doesn't just go for two or three minutes, it goes all night.

Related to this is another concept I like to think of as 'the illusion of music'. A big part of what makes music work is the way that it sticks to certain rules and patterns. Yes, that's right, even the music genres with the most rebellious personalities stick to *some* rules.

The human brain *loves* patterns. When it hears patterns in rhythm, harmony and arrangement, it locks onto them and gets hooked trying to guess what comes next. A few seconds of music is all it takes to create an illusion of music and plunge your imagination right into it (not to mention your body). The greatest thing about dance music is the way it keeps this illusion going all night. The rhythmic rules in particular are continuous between songs, passing the illusion from track to track. The illusion is subtly changed throughout the night, according to the tracks the DJ selects.

Destroying this illusion by breaking these rules is the fastest way to evaporate your dance floor. You may recognise the feeling from a time you've done a badly beat-mixed transition during a set (yep—we've all done them). The beats from one track are splattered all over the other, and suddenly everyone is pulled out of the music, blinking and looking around the dance floor, unsure of where the beat is, and how to dance to it. That's usually their cue to go to the bar.

When we talk about rhythm and key, you'll develop an understanding of how you can follow these rules, and make sure our different musical elements follow the same rules. This way, you can learn to keep the illusion of music going, as the original artists of each song intended.

HOW TO USE THIS BOOK

It is my hope that this book will help you to not only understand the principles behind how, why and when to use mashups, but also to learn a set of techniques that can guide you through almost any technical problem that presents itself to you. As you read, I encourage you to have Ableton Live open and to try each technique as you learn it. There is also a companion website with a ton of videos to help you understand the techniques, see them being done and hear why they work (visit 'makegreatmusicmashups.com'). As a DJ, you're likely to learn best by hearing examples in action, so each time you see a reference to a video, go ahead and open it. Other than that, the book is ordered in sequence so that it is easy for you to come back to at any time—from necessary equipment and software, preparation through to execution and house-keeping.

In the next chapter we'll look at the components of a mashup.

CHAPTER 2

The Components of a Mashup

The next few chapters will be all about the theory behind what goes into mashups. Many new mashup artists are intimidated by the vast world of musical theory and decide to stay away. That's fair enough, because there's a lot of information out there that won't be relevant for them. Rather than go into complex musical theory, I'll be simply showing you the basics of what you need to know to keep the illusion of the music alive and to get your mashups *sounding good*.

In later chapters, we'll be opening up Ableton Live and actually going through techniques, but for now let's talk about the ingredients we need for our mashups.

These are the components you will be using in your mashups:

- main track (one or more);
- breakdown-only tracks/classics/non-dance songs;
- vocals (sung acapellas, fills, raps, chants);
- musical riffs/sections;
- added drums and effects.

THE MAIN TRACK

This is the part that contains the big drum sections and **intro/outro**. It makes up the loudest and most hard-hitting part, or 'dance' part of the mashup. In other words, it's where the beats come in (more recently referred to as the '**drop**').

In most mashups I've either made or come across, this is the element where DJs choose to go looking for a brand new track. Therefore, the audience is often being exposed to it for the first time, while the other parts are there to give them some familiarity. Here's an example; you are putting together a mashup for a commercial venue, and you've just purchased a great new big-room track you'd love to play. However, the track is unknown to the audience, and it's a little too high-energy for the venue. In order to make it work, you give the audience a breakdown using a current radio track to get them comfortable, before transitioning to 30–45 seconds of the *main track*. This way you're able to give them something familiar to relate to, before using a smooth transition to ease them into the energy of the new track. In this example, the function of the mashup is to prime the audience for something new and fresh by tempering it with something they already know. But this doesn't always need to be the case.

Some older 'main tracks' are too good to just throw away and stop playing. This is particularly true if it is a track you have produced, in which case the audience is expecting to hear it anyway. Therefore, it's a good idea to use mashups to revitalise those songs you want to continue to play, so that they don't become boring.

Wherever you source your music from, it's important to get it at a decent audio quality—at least 16 bit wave files (.wav extension, CD quality) or a 320**kbps** (kilobyte per second) **mp3**. Often, you can only find acapellas in a less than adequate quality, but the main track is very important to get at good quality because it is sonically dense and will be playing on its own at some point. For newer tracks this is no problem, but finding older music at a high-quality bitrate is often difficult.

The way you source your main tracks—or your dance music collection—is really down to what kind of DJ you are. If you're starting out, the most popular place to get new dance music tracks (at the time of writing) is Beatport.com. They have the largest range of dance music, and in addition to being a portal for purchasing the music itself, their top-100 structure makes it easy to see what DJs are buying within any given genre. Therefore, it is also a useful place to preview the music and get a sense for what DJs are playing (or at least buying) globally. Again, where you source your music from is highly personal, and there are a lot of DJs and fans who heavily discourage the idea of just playing what's at the top of the Beatport charts in order to keep dance music from becoming generic and commercialised. So rather than get into that debate, I'll leave you to decide whether or not Beatport is right for you in terms of discovering your music.

Online music blogs are also a good source for discovering the music you want to use. If you find a few blogs that post music you associate with in your sets, it helps you find the music you don't necessarily find at the top of the lists on Beatport, as well as letting you know what is coming up over the next few months. Some music is even given away for free in return for social media follows or likes.

Record labels also do their own promotional send-outs, so it's worth trying to get on their list if they produce a certain sound that resonates with your set. Labels like to see that you are somebody worth sending out their music to for free, so it's a good idea to have some of your best sets up on Soundcloud, or some proof that you have a following before you contact them. If you don't have much to show them yet—mashups may be your key to getting there.

For older music, you can rely on more traditional music distributors. Just be wary though that places such as iTunes generally use heavy file compression on their audio files, and you shouldn't purchase them here unless the music is available in Apple Lossless (which is CD quality audio). When looking for old music, often your only option is to purchase the physical CD or vinyl on eBay or Amazon. It's still worth checking Beatport or other dance music distributors to see whether the music has been re-released for the digital era.

BREAKDOWN TRACKS, CLASSICS AND NON-DANCE SONGS

This is where you get creative. In between sections with big drums, you can choose not to use the main track at all. You can choose to drop to a song in a completely different genre, or even speed, as long as you're able to transition in a way that still makes sense. You could choose anything from the breakdown of a new or recent club track, to a classic house or trance breakdown, an old rock song, a pop track, or an R&B favourite.

What you choose to use as your breakdown track is very dependent on the 'sound' you want to create for yourself, and your knowledge of the crowd. This is the component that allows you as a DJ to take an audience wherever you want to. You can sidestep them into the world of radio music, back in time to a classic track or even into their favourite movie motifs.

If you need some inspiration, listen to a few sets from your favourite DJs. Genre and era are no barrier, so long as you can find other elements to successfully pair it with for any given mashup. In fact, some might say that the act of choosing a breakdown element is the most creative part of mashups, because unlike current club music, there are no charts to tell you what to use (with the exception of pop music). It all comes down to your experiences, your musical history, and your interpretation of what a venue or crowd might relate to. When thinking ahead for a specific show, think about the age of the patrons, the other acts they've had there, or whether the club is known for a particular sound. Similar to DJing in general, try to manage their expectations by balancing your breakdown choices between what is 'safe' and what is surprising.

VOCALS

When dealing with a vocal, DJs commonly use the term acapella. Traditionally, the word acapella (from the Italian 'a cappella', meaning 'in the manner of the chapel') means a piece that is performed only by voices, such as a choir without instrumental accompaniment. In the world of DJing and production, an acapella refers to an audio file with nothing but a song's vocal performance on it. The internet is swarming with them. As discussed earlier, house music has known a long history of records being intentionally released with an acapella version. DJs now have access to an incredible range of vocal-only versions, and therefore an incredible potential for great mashups. Record labels such as Defected, understanding the value of these little gems, actually releases regular compilations of acapellas taken from their biggest songs.

It doesn't just stop at dance music. Acapellas exist for songs of all styles. Often when music is being mixed down, vocals are **exported** separately in preparation for producers to use in remixes. Officially released acapellas have become less popular in modern times, so often you won't be able to find them for new songs. However, there are still some tricks you can use to incorporate their vocals and exclude much, if not all, of their other elements. We'll get into that later, once you are introduced to Live. In the meantime, a simple Google search for acapellas will bring up plenty of sources; my favourite for older tracks is www.acapellas4u.co.uk, which includes thousands of dance and pop/R&B vocals. When looking for newer vocals, a web search for the term 'acapella' and the name of the song should let you know if one exists pretty quickly.

Vocals are one of the most valuable tools in mashups. As a single sound rather than a whole track, they are far more flexible. It is much easier to blend an acapella with a full track than to try to blend two full tracks together. They are also far more usable because the human ear is a lot more forgiving on imperfections in the voice than it is on other sounds. By their very nature, vocals demand attention. For people who are only just getting into clubbing, vocals supply something lyrical to listen to, something they are used to from listening to radio and personal music collections.

MUSICAL RIFFS/SECTIONS

A rarer element in mashups, these musical riffs can provide big moments. If done well, they can be used over full tracks or break sections. Similar to an acapella, they are a section of audio that contains nothing but a single musical riff. Sometimes these are sampled from records that drop to a famous lead sound, or sometimes you can recreate them yourself. Often all you need is eight seconds of a **solo** instrument at the start of a song and you can sample it (borrow some audio from a track and save to a new file). Then you are in business.

Some classic examples commonly used by DJs are the iconic synth riff from Da Hool—'Meet Her At The Love Parade', the synth strings from Tiesto's version of 'Adagio For Strings' or the charismatic sawtooth melody from Pendulum's 'The Island'.

Musical riff samples can even be **loops** from classic rock/pop songs, such as the bass-line that appears on its own at the start of 'Seven Nation Army' by The White Stripes, the piano from Coldplay's 'Clocks', or the famous string melody that opens 'Bittersweet Symphony' by The Verve. Having access to a single element like this makes it far more versatile; it can be used almost like a remix element, allowing you to chop and edit it how you want without other instruments from the song causing problems. Truly effective mashups can be made when you allow a song to play as-normal for a breakdown or main track element, but also sample a single riff from the song and use it throughout the mashup to help the whole piece blend together.

ADDED DRUMS AND EFFECTS

When combining two or more pieces of music together, you don't always have everything you need. Sometimes, no matter how well you edit and arrange your pieces, there will be dips in energy, small sections that don't join properly, or a need for a specific sound effect to help let the audience know what is coming next. In this case it is necessary to add your own sounds in order to create a feeling of continuity, and to replace sounds that have been lost as a result of your edits. In a best-case scenario, you can steal little loops or individual sounds from one of your elements, to keep the sonic flavour of the music in keeping with the rest of the mashup. These sounds can include a boom or crash that appears on its own at the end of a track, a section of the intro or outro utilised as an added drum loop, or even a single drum hit like a kick or **snare**, provided they appear on their own.

Other times, it is impossible to sample the part you want, and you need to have your own sounds ready to help you fill the gaps. It's important to have a library of samples available to you that can fit any situation. Vengeance Samples are a popular yet genre-varied source for sounds such as drum loops, individual drum hits, sweeping sounds (upsweeps and downsweeps), sirens and risers, drum-rolls, impact sounds and more. There are many websites out there that sell sounds for producers though, you simply browse what you want, purchase sample packs and download them. There is also a collection of sounds available from the companion website for you to use ('makegreatmusicmashups.com').

Later, you'll learn more about what each of these elements brings to the party, and about how each of them must be treated differently while you work with them in Ableton Live. It's also crucial to correctly label and organise everything you use so that it's easy to find the next time you want to use it in a mashup. We'll discuss a watertight system of file organisation later on. But first, let's look at how to choose the right mix of elements for your mashup.

CHAPTER 3

The Right Mix of Elements

Let's consider that we have main tracks, big breakdown moments, vocals, some effects put aside and maybe even some musical riffs. What do we do with them now?

> To me it's about taste. Bottom line—choose combinations that genuinely work—it's like cooking. Yes, sympathetic keys are vital, as is good editing technique. But none of that really matters if your main combination is a bad choice to begin with.
>
> (James Ash, DJ/Producer)

> Mashups aren't restricted to any genre, and as long as the concept is there, you will always have a way to innovate your set, and stay in tune with any current trends.
>
> (DJ Trim, Australian Club DJ)

What is more important to a great mashup than using the right techniques is picking the right elements. Let's use a really bad example first. If you were to take a country music song, and throw a death metal screaming vocal over the top, you are probably going to confuse and irritate both your country fans and your death metal fans. What would be worse is if they were so far apart in rhythm and key that they simply could not be made to work together.

Big artists that sit halfway between DJs and live music acts, such as Daft Punk, Justice, Porter Robinson and Madeon are famous for making their live sets an ocean of mashed-up ideas. Their sets will position a vocal from one song over the bass-line of another, merging from one song to the next, chopping and changing through different elements from each of their songs. This is a special treat for their biggest fans, because they know which element belongs to which song. But it makes perfect sense when you think about it; an artist will often write songs that have similar rhythms, similar speed, similar arrangements, similar emotions, even similar types of sounds. Why shouldn't they work together? This kind of logical mingling gives us a clue into what technically makes a good musical pairing. What you want to find in your pairs are tracks that have some common traits, even if they seem different in other ways.

There are a number of things we can listen for to help us find these traits.

Energy level: When we talk about the energy level of a track, we're talking about the power or excitement in it. A quick listening comparison between two pieces of music should give you a pretty immediate reaction as to whether one is too highly-energised for the other. Whilst mashups can be a great way to transition from one energy level to another, they're going to be harder to manage as the difference between your tracks becomes too great, and the audience will find the change too sudden to accept. If you are already DJing, you will be quite used to this concept, knowing just how much of a change you can make with each new song.

Energy comes in all different forms too. Sometimes it is determined by the speed of the track; trance at 140 beats per minute is likely to sound more energetic than R&B at 90 beats per minute literally because of the greater number of musical events that hit the listener with each passing second. In this case it's important to remember that if two songs with different speeds are brought together to one **tempo**, their energy will rise or lower depending on whether they move up or down in speed.

Energy is also defined by the types of instruments used in the production. A big wailing electric guitar riff or a savagely processed electro lead will create more intensity than a subby, plodding bass-line or a soft acoustic guitar. The most effective test you can run when comparing two potential pieces is to listen to a snippet of each in quick succession using the music program you use to run your DJ library, or just a regular music player such as iTunes. If one song seems to leap out in front of the other, you know they may be hard to match. Even vocals make a difference—the party-rocking rapping of Fatman Scoop or Lil Jon are going to incite more excitement than the smooth emotional melodies of Sarah MacLachlan or Coldplay. What is interesting about this particular comparison is that it illustrates the very different emotions each example contributes. Which brings us to the next attribute to listen for . . .

Feel/Emotion: The closer the 'feel' of the tracks are to each other, the easier it's going to be to blend them. Some vocals, for instance, are too ethereal or soft to effectively sit over the top of a hard-energy, aggressive backing. No matter how well they are blended, they are just not going to work. In the beginning, new mashers should try to stick to styles that work seamlessly with each other, as it reduces the difficulty and the work involved in trying to make them blend. However, don't be afraid to experiment later on, because pushing the boundaries of what you think will work can create some interesting and often surprising discoveries. You also want to think about whether the elements you are blending sit close together in terms of fun/serious, happy/sad, cool/over-emotional. For instance, if you're playing an intensely emotional trance track and a big, fun party vocal comes in over the top, the crowd will probably feel very confused, not knowing which emotion to follow. The cleverest mashup combinations come from matching tracks that have some similarities, yet some differences. My favourite example is rock and electro, both of which are moderately fast and aggressive, yet are opposite sides of the spectrum in terms of their electronic/ live feel. This grounds the audience in one way, keeping the illusion of music alive, but still gives them an interesting change of direction.

Here are some aspects to think about when looking for common ground between elements:

- excitement: relaxed/high
- emotion: neutral/highly emotional
- era: classic/modern
- electronic/'real'
- fun or serious
- happy or sad
- aggressive or passive
- masculine/feminine
- cool or alternative/pop and mainstream
- spacious and open/busy and full.

Time-slot: Another great way to think about whether a pair of elements are appropriate together is to imagine the finished mashup and picture yourself playing it out. If you can easily see yourself playing it at a specific time of night without fear (i.e. early set, late set etc.), then that is a good sign. If, on the other hand, you can see yourself playing one of the elements in an early part of the night, but wouldn't dare play the other element until 3 o'clock in the morning, you might want to save one of those elements for a mashup in a more suitable time-slot. This is where DJ experience comes in, but if you are new to DJing, don't fear. As you make mashups you will develop your understanding of the links between music, which will help you better understand DJing as well.

Key Clashing: Pretty much all pieces of music feel like they belong to, or 'point to' a specific musical note. The keys of your two musical pieces must be reasonably close, enough that you can use **pitch** correction to match one to the other. If you test two elements together only to discover that some musical notes clash badly with each other, you are likely trying to blend elements that are incompatible. Stay away from these combinations! We'll go into how to identify the key of a song and what technically prevents them from blending together when we talk about keys in the next chapter.

Lastly, and more critically than anything else . . .! Just use your ears. It's important to get very used to using your music player as an element testing tool. Switching from one piece of music to another will give you a hint of what the dance floor will feel on their first listen, so learn to pay attention to your first emotional reaction. If you know in your gut that there's some link between two elements, throw them in a sequence and try it out. You'll know pretty quickly if it's wrong, and figuring out *why* is a great lesson that will improve your skills for next time.

> With the technologies we all have access to today it is easy for anybody to make mashups—which is great but also means that there is now an abundance of mashups out there that are a bit questionable and, in some

cases, downright wrong sounding . . . Often the keys are not gelling well, grooves fight or the new vocal/music section etc. just sounds horrid as it's been ridiculously over time-stretched or re-pitched. I think having a well thought-out blending of elements that is properly executed is the most important thing.

(Kam Denny, DJ Producer)

CHAPTER 4

Mashup Theory

I promised you we weren't going to get into ridiculous amounts of musical theory, as there's so much theory that won't apply to creating mashups. So, I'll share with you just what you need to make sure yours sound great. The main areas we'll cover are key and tempo.

KEY

> The two (or more) tracks you're mashing need to be musically cohesive. Nothing worse than an out of key mashup in my honest opinion.
>
> (Ivan Gough, DJ/Producer)

Key is an incredibly important aspect in mashups; nine times out of ten, when a DJ tries their hand at mashups, the greatest problem they struggle with is firstly figuring out what key their elements are in, and then figuring out how far and in which direction to move one of their elements to match the other. Or worse, they just hope for the best and throw them together anyway!

If you listen to a great DJ set with plenty of mashups, the best ones always sound smooth, as if the elements *belong* together. This is a crucial concept. Sometimes mashups feel so smooth that you can't even tell if it *is* a mashup. The reason for this is that the elements have been picked for their musical relationship with each other, and all belong to the same family of musical notes. Great DJs even pay attention to key when selecting which song to play next, in order to help their whole DJ set feel like one single epic piece of music.

> The most important thing about a mashup for me is when you can't even tell it's two separate songs. It's so well-crafted and matched that it just sounds like an original song.
>
> (Pierce Fulton, DJ/Producer)

In music theory, key is one of the first concepts you learn. Every piece of music is in a particular key, and that key determines, among other things, which note on the keyboard feels like the 'home' or centre of gravity of the song. The melodies will tend to keep coming back to this central note, revolving around it, always implying it and trying to come back to it. Matching together musical elements that share the same home is how you find combinations that feel at home *with each other*. You may have heard this 'home' note also referred to as the 'root note'. Some styles of music make it easier than others to find their root note and subsequent key. Dance music is usually easier to figure out—with very limited use of notes, most hitting on the root note or notes very closely related to it. Even when you don't consciously understand what this means, the subconscious brain *does* recognise it, and when you hear a piece of music, it quite quickly finds the key, allowing you to enter the illusion of the music. Unfortunately—like a lot of concepts in music—the subconscious brain doesn't like sharing this information with our conscious brain, and we have to dig a little deeper to understand what we are hearing and why it feels the way it does. Thanks a lot, brain.

There are only 12 notes that exist in modern western music. If you're not used to musical notation, the way they are organised can certainly seem a little strange. The easiest way to become familiar with them is by looking at their physical arrangement on a keyboard or piano: seven of the notes are named A through to G (the white notes), and five more (the black notes) are referred to as the **sharp** or **flat** versions of these original notes. Figure 4.1 shows what they look like on a keyboard.

FIGURE 4.1

Note: Black notes on a piano can be represented as either *sharps* (a semitone higher than their natural counterpart) or *flats* (a semitone lower). In traditional music theory, how you represent them would change depending on the **scale** you're using, but for the purpose of simplicity, I'll simply be representing all black notes as the *sharp* versions of the key just below them. I suggest you also stick to using just sharps when noting down keys or labelling your mashups, as this will keep things much better organised later on.

FIGURE 4.2

You'll notice in Figure 4.2 that once we get to G#, the notes start to repeat, beginning from A again. This is because each different letter note appears multiple times across the keyboard. We call the distances between the different versions of the same note **octaves**, in that one 'A' is one octave away from the next 'A' up on the keyboard. They both sound like an 'A', only one will sound higher in pitch than the other. The further to the right you go up the keyboard, the higher the pitch of the sound. To put it simply, the bass-line notes occupy the low pitches down the bottom of the keyboard, whilst high-pitched instruments occupy the top notes.

Let's delve into relationship between the 12 notes a little more by pretending we wrote a short piece of music that used all 12. What would happen is that the human ear would be confused by such a bizarre pattern of notes, and wouldn't be able to figure out which note is the root note of the track. This is why kids banging their hands on a piano jumps out straight away as sounding 'unmusical' (even if it sounds hilarious). It's because our brain doesn't recognise the pattern, and cannot find the root note.

In classical forms of music, this root note is usually quite easy to find, because it finishes its primary melody on that note. Back in the day, classical music made a big deal of 'coming home' or 'coming to rest' at its root note at the end of a piece, giving it a feeling of finality and stability.

Sounds straightforward enough, right? Bringing it back to mashups; if you buy a shiny new track in the key of D that you want to put a vocal over, the best place to start looking is for vocals in the key of D. Is it that easy? Not quite, unfortunately. Key doesn't just describe which note is the root; it also describes the type of 'scale' it uses.

All of the notes present in a piece of music other than the root note are determined by its scale. Though there are 12 notes available, a single piece of music usually only uses around seven of them. This combination of notes make up the scale; a specific selection of notes that the music *operates within*, and which give it a specific feeling and tone. Think of the scale of the song as being an instruction manual that tells the song where its notes must live. If a song breaks these rules by placing a note outside of its specific scale, the note will immediately jump out at the listener, upsetting the subconscious. What this does is break the illusion of the song for the listener. Suddenly, the emotion that the song has successfully created falls flat on its face. This is why our little friends slamming their hands on the piano don't sound as musical as our more celebrated artists—musicians pick a scale and stick to it.

So, back to what this means for mashups. In addition to being in the same key, the two elements you want to blend must also be in the same *scale*, or many of their notes will have an incredibly high likelihood of clashing with each other.

Here's where it can get a little bit confusing. The important thing to remember about scales is that regardless of what root note they begin on, each scale type uses a specific *sequence* of pitch distances between each note, made up of **semitones** and **tones** (Figure 4.3). A semitone is one note distance, describing two notes that are right next to each other on the keyboard; for example the interval between C and C#. A tone is two note distances, for example the interval between C and D, which skips C#. This sequence determines which notes are acceptable for use in a song, and which ones must never be used. Note: North-Americans call semitones *half-steps*, and tones *whole-steps*.

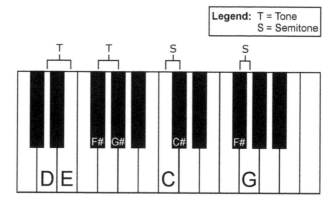

FIGURE 4.3

Fortunately for us, most modern music can be classified as using one of two types of scale. We mentioned them before; the major and the minor scale. The most common scale for dance tracks is the *minor* scale, because this particular sequence of notes gives music a more 'serious' or 'cool' feeling. Outside the dance realm, the minor scale is commonly thought of as having a 'sad' feel to it. Beginning from the root note, it always progresses in the following pattern, ascending upwards in pitch:

Root, Tone, Semitone, Tone, Tone, Semitone, Tone, Tone (reaching the root again at the top).

Figure 4.4 and 4.5 are two examples of what a minor scale looks like on a keyboard.

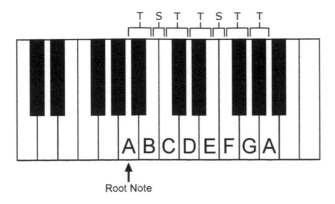

FIGURE 4.4

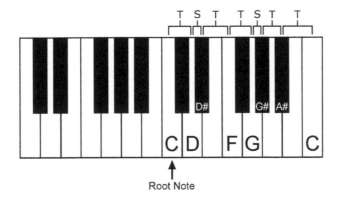

FIGURE 4.5

For those that are familiar with music theory, you will recognise that the scale pictured is a *natural minor* rather than a *harmonic minor* (G rather than G#). Whilst the traditional harmonic minor uses the sharp 7th, almost all dance music uses a natural 7th as it takes it from being a 'tragic' feeling scale to a 'cool' feeling scale.

The minor scale encompasses most dance music. It would be a good idea for you to jump on a keyboard and play the notes you see in the examples above from bottom to top. This will train your ear and help you to remember what minor keys sound like.

However, many vocals and non-dance tracks you will come across will be in a major key, which gives them a more happy, fun, or uplifting feeling (Figure 4.6, 4.7). They can be quite common in your breakdown tracks, and therefore it is important to understand them as well.

The sequence for a major scale is: Root, Tone, Tone, Semitone, Tone, Tone, Tone, Semitone (reaching the root again).

Compare the A Major scale in Figure 4.6 with the A Minor scale from Figure 4.4. Even though the root of the key 'A' is the same, the scale is different, because some of the other notes are changed (the 3rd, 6th and 7th).

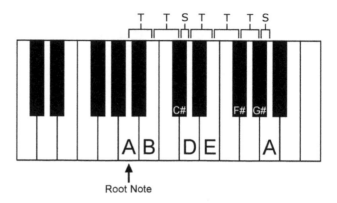

FIGURE 4.6

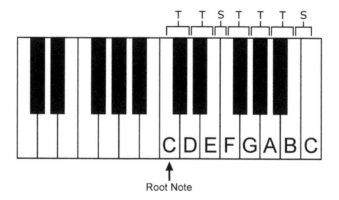

FIGURE 4.7

Have a play through the major scale if you can, and notice how it feels different to the minor scale.

The reason it is important to use two pieces of music in the same key in a mashup is to enforce that restriction on the notes; to maintain the same scale while the two elements blend together. It keeps the illusion of the music going, even as the instruments, lyrics and melodies are being passed from one to the other, or simply running together simultaneously.

There are more subtly altered versions of the regular minor or major scales that exist, and occasionally tracks that are neutral, with no obvious major or minor notation present—but for the purpose of mastering mashup theory, these two are currently all you need to know about.

FIGURING OUT THE KEY OF AN ELEMENT

So, if we've just found a new track or vocal to use, how do we figure out what key it's in?

This is one of the hardest things to learn when first approaching mashups. Fortunately, dance music makes this a little easier by mostly being composed in *minor* keys. But it's always important to check and be sure. Playing an element you *aren't* sure about right after one you *are* sure about is a good way to test this.

When you start out, it's a good idea to come up with a list of reference tracks to test against. As you gradually begin figuring out the key of your music collection, create a playlist of 12 songs in the minor scales from the key of A through to G#, and then 12 songs in A major to G# major. Then whenever you aren't sure what key a new track is in, you can test it against your references to be sure you've found the right key.

Outside the realm of dance, you need to be more cautious. Vocals and musical riffs taken from pop or rock songs have just as much chance of being in a *major* key.

Just a word of warning. If you plan on using music released before the year 2000, you should be on the lookout for slightly off-key tracks, particularly in the dance genre. Many digital recordings of songs from that era are actually sampled from vinyl using techniques that do not always render at exactly the right speed. It's not unusual to find classic records that are a tiny bit faster or slower than they were originally produced, and as such they have moved up or down a little in pitch. If you try to play a piano or synth note over the top to find its root note, it will seem to sit somewhere *between* two keys. It's important to use Ableton Live or another program to push them back into the key they are supposed to be in before you can blend them with anything. The most sure-fire method is to load them into Live and play them up against a long synth note in the key you think the track should be in, then change the pitch of the track until it sounds perfectly in tune with the note. Though this can be tricky to do by ear, it is a very important skill to learn, and you will get better as you work on more mashups. You can use the frequency analysers that come in many music production programs to help you if you get really stuck.

Interestingly, the odd occurrence of slightly off-key music also takes place in the digital era, but very rarely. When Avicii released his 2011 monster track 'Levels', the key was a pitch roughly 20 cents (20 hundredths of a semitone) higher than C# minor. If you listen to the sample on which he based the track, Etta James's 'Something's Got A Hold On Me', you can hear that he has pitched his instruments around the sample, in order to preserve the sonic integrity of the original sample. When DJs tried to mash it up, however, it resulted in an explosion of out-of-tune mashups in the club scene because they didn't correct their other elements to account for the 20 cent difference between Levels and the true pitch of C#.

> On a sidenote, the online dance music store Beaport.com includes key information with their releases, but be very cautious because during the publishing process it is often keyed incorrectly. Make sure you check it yourself. The key-labelling system they use is also unable to account for off-key music like the example above.

If you don't have much experience with musical keys, I can fully appreciate that you're dreading applying what you've just learned in order to find the key or music you want to use. Luckily, there are a couple of techniques you can use to help yourself out. The first is that I would suggest using the software 'Mixed In Key'.

MIXED IN KEY

> Mixed In Key is basically a circle of fifths for the every day person. You drag and drop songs into the application and it analyzes them and tags them with a key of a letter and a number. It's designed to be an intuitive way to harmonically mix by taking your number/letter combo and always matching with the one above or below or the relative minor/major.
>
> (Pierce Fulton, US DJ/Producer)

If you haven't heard of it already, a MacOS and Windows compatible software called *Mixed In Key* is a fantastic tool that was developed to help DJs **mix** their records in key during shows, making for more natural musical transitions. Many DJs use it, and are able to have their mixes blend even more smoothly by connecting tracks that are in the same key and helping the set glue together.

Of course, this is a priceless tool when it comes to mashups if we have trouble picking the key of a track. If you are interested, have a look on YouTube to see some examples of DJs using the software to plan out their sets in musically cohesive ways.

When you import a track into the software, it analyses it by studying all of the **frequency** information and finding patterns in the notes that happen to be used throughout the song (sound familiar?). Then, it spits out two values: what the root note of the song is, and whether the scale is major or minor (also sound familiar?).

Figure 4.8 shows how the values it uses represent musical keys:

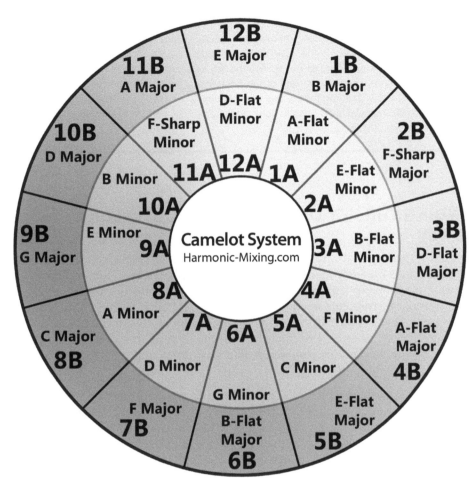

FIGURE 4.8
Mixed In Key's Harmonic Camelot Wheel

www.mixedinkey.com

This is a good illustration to have stuck to the wall near your computer as you start out with mashups (see the website for a copy). The program uses the code of a letter (major or minor) and a number (the twelve keys) to describe the musical content of the track. The program also adds this code to the end of the filename of the file it analyses, so that DJs can see the compatibility of tracks while they're playing, or even group their songs by key.

Whilst many DJs let the software label the track filenames with the program's own jargon: '3A, 11B, 5A' etc., I encourage you to keep a note of the actual musical keys, so that you don't have to convert things later when you're trying to find good match-ups. Also, if you later decide to get into music production, you'll need to understand keys, not letter and number codes.

A word of warning—many vocals or small solo elements only use a small collection of the available notes within the scale of the song they live in. Therefore, if you are attempting to use Mixed In Key to find the key of a vocal, or short musical riff, I recommend that you find a version of the *full track* to import into the program for analysis, as a solo vocal or melody has less frequency information for the software to analyse. You can then go back to your acapella or solo instrument and add the key information to the filename.

FINDING THE KEY OF A SONG WITHOUT MIXED IN KEY

If you'd rather figure it out for yourself, the first thing you'll need to do is to get your hands on a keyboard or an app that can play notes. There's plenty of freeware programs or smartphone apps that do it. Otherwise you can use one of the Virtual Instruments in Ableton Live such as *Analog*, since you are going to be using Live anyway. If the element you are trying to figure out is a solo vocal (acapella), I suggest listening to the original song in full while figuring out the key, as solo vocals sometimes skip past notes that help identify their key.

Even if you use Mixed In Key, it's a good idea to try the following method to help you better understand and identify key. I would strongly suggest starting with some dance music, as it is usually extremely obvious with its key information. Because repetition and power are the name of the game in dance music, it tends to favour long sections that remain firmly attached to the root note, without complicated musical changes. If you get the hang of it, you can move on to rock, pop and R&B, which are still reasonably simplistic but do move around a bit, using chord progressions to temporarily pull away from/settle back down to the root note.

To find the key, load up your keyboard or app. Next, listen through your element in a music player. While it is playing, see if you can pay attention to times that the track 'comes to rest' or feels as if it is moving from tension to 'release'. During times that the track feels as if it comes to this resting point, see if you can hum the note that the **bass**, low-sounding instrument is playing. If you have got it right, this will be the root note. As I mentioned, this is the trickiest part, and if you're not used to it, may require some practice before you become quick at it.

Then, while still humming, jump on your keyboard; playing your way up the keys, one note at a time (Figure 4.9). Start from anywhere you like. If it's easier, you can pause your song, but just make sure you don't lose your humming pitch. Continue up the keyboard until you find the note that matches the one you're humming, and write it down.

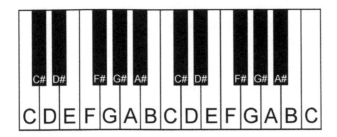

FIGURE 4.9

If you have trouble finding the 'resting point' of the song, try the following method instead. Listen to the song for 15 seconds or so, and then while it is still playing, try humming the first note you think it *might* be, and continue to hum it, listening to whether it feels good over the *whole song* (no matter what chord changes occur underneath).

Most music creates the feeling of movement and progression by creating *chord changes*. A chord is a group of two or more notes played together, and chord changes occur by having the bass, keyboards or other instruments move around on different chords (with each other) in a repetitious chord progression. They help songs feel like they are telling stories, but by keeping the chord notes limited to the notes within the scale, the song doesn't break the illusion of the song. In fact, the song usually feels like it repeatedly keeps coming back 'home' at some point in the progression, which is what creates the constant pushing and pulling of tension and release.

Most of the time, when you hum, your brain will give you the root note. First, as mentioned, test by humming it continuously over the song to check that it always feels good. If it doesn't, try another note until you feel confident in how comfortable the note feels when hummed over the song *regardless of any chord changes in the piece.*

Whether you used the first or second method, we can perform a check to make sure the note you found is *definitely* the root note. Occasionally, the note you found will be very close because it is 'harmonically related', i.e. very close to it on the Camelot wheel. But it's not quite the root note.

To test it, try playing a major or minor scale starting from what you think you have identified as the root note. This test will not only determine that the root note is correct, but will tell you whether the song is in a major or minor key.

While the song or element is still playing, get out your keyboard or app, and locate your root note again. Then, using the sequence for a *minor scale*, begin playing each note in the scale.

> Minor scale: Root, Tone, Semitone, Tone, Tone, Semitone, Tone, Tone (reaching the root again at the top).

If the notes you're playing as you go feel like they make sense with the music, then you've found your scale. *All* of the notes should feel right. If, for example, your root note was D, your song or element is in D minor.

If, however, it sounded terrible and pulls you out of the illusion of the music, try going back to the root note on the keyboard and playing the sequence of notes that make up the *major* scale. If all of the notes make sense over the music, you have found your scale.

> Major: Root, Tone, Tone, Semitone, Tone, Tone, Tone, Semitone (reaching the root again).

It can be difficult trying to figure out the scale using a sequence of tones and semitones, so I suggest you go to the website for a free downloadable pdf with illustrations of all the major and minor scales as seen on a keyboard. It's a handy one to print out and keep near your computer. ('makegreatmusicmashups.com')

If some of the notes still sound horrible and you hear clashing between your keyboard and the music, then you have most likely located the wrong root note. In this case, go back to the start of this method and try again with another note. Don't worry if you struggle with this at first, this is incredibly difficult when you are new to it. Understanding keys intuitively doesn't happen overnight—but with practice you will eventually be able to hum the right note first try. You'll also be able to pick the major or minor by feel, rather than needing to test it by playing the scale.

Of course, if you really struggle on certain tracks, there's always Mixed In Key to help you out.

Lastly, let me share one thought on organisation. I learned early on after having to figure out many keys and tempi again and again that the best practice is to label a file every time you figure out one of these attributes. This is particularly important with the **BPM** (speed) of your acapellas, as tempo detection algorithms struggle to correctly analyse audio that doesn't have drums in it. I can't even begin to guess how many times I had to figure out the Acapella tempo for Camille Yarbrough's 'Take Yo' Praise' (known for the vocal in Fatboy Slim's Praise You), and cursed myself for not writing it down the previous time I used it. Do yourself a favour . . . add key and tempo details to the filenames of your wavs and mp3s!

DECIDING THE KEY OF YOUR MASHUP

What happens if we have a brilliant idea for a mashup that includes a main track and a vocal but they're not quite in key? For example, we have a main track in C minor, and a vocal that fits it perfectly in terms of mood and set-time, but it's in D minor, only a short pitch difference away? First, let me explain what happens to audio when you want to re-pitch something.

Audio is made up of many tiny little **samples** that describe the position of a speaker cone at any given moment. If the signal stays flat, the speaker cone doesn't move, and you get no sound. When you play an audio signal with some noteworthy **waveform** activity going on, it will cause the speaker cone to move and produce sound.

In order to increase the pitch of an audio signal, your software must squish the waves within the waveform closer together, so that they become a higher frequency, therefore being interpreted by our ears as higher pitched. Though this does increase the pitch of sound, it also means that it is being played back *faster*.

This is the major problem we have in shifting the pitch of elements—trying to keep their speed the same.

Conversely, if we try to make music faster or slower, this means that its pitch will move as well, even when we don't want it to.

One of the major advantages of Ableton Live is its superior and intuitive ability to 'Warp'—a process that time-stretches audio so that you can shift one attribute without changing the other. I'll show you how we do this when we get to Chapter 6. In the meantime, just remember that Warping, like any form of time-stretching, has to essentially sacrifice the quality of the audio to perform the change, so there's a limit to how much you can shift an element before the problems outweigh the advantages.

Back to our brilliant mashup idea that involves two tracks that are two semitones apart. What do we do? In some instances the two elements are close enough that we can bend one or more of them to reach a common pitch. But we have to make a choice about which ones to leave as they are, which ones to pitch-bend, and how far to bend them. We will be required to pick a key that we want the mashup to appear in, and then bend all the elements to that key. Some element types absolutely need to stay as close to their original pitch as possible—others can move a lot more freely without noticeable damage. Here's a list to help you prioritise which elements *must stay as close as possible to their original pitch*. Items at the top must be preserved as much as possible, items at the bottom are more forgiving to pitch-shifting.

1. *Main track*—try not to shift this because pitching will destroy your drums and create very noticeable audio glitches.
2. *Breakdown track*—this is less of a priority because there will be little or no drums.

3. *Musical riffs*—these can bend even further in pitch because the sounds are far less complicated.
4. *Vocals*—these can bend the furthest, because they are only a single, simple element.

You might notice I didn't include one last element on that list—effects and extra drum/build elements. Fortunately, these have no tonal element to them, so they never actually need to be pitch altered anyway.

As a general rule, try to never move the pitch of the main track unless it's absolutely necessary. I would happily move a vocal up by two semitones to avoid moving the main track. However, any more than two would be pushing it. For an example where the two elements are three semitones apart, I would pitch the vocal up by two semitones, and pitch the main track down by one, and if it sounded terrible, I'd ditch the idea.

Of course, the important thing here is to use your ears. Some vocals sound perfectly fine when pushed up in pitch by two semitones, but terrible when pushed down by even one. Some full tracks may occasionally sound fine bent one semitone off their natural pitch, yet in others you'll notice horrible distortions in the kick and bass that completely kill the vibe on the dance floor. Trust your ears and remember that mashups are about delivering a moment, and if you deliver them a pitch-bent vocal that sounds like a chipmunk, people are going to think you're nuts. Always ask yourself whether the benefits are worth the side effects.

We'll cover more on ways to alter pitch successfully in Ableton Live in Chapter 6.

TEMPO AND RHYTHM

If you aren't familiar with the term *tempo* (from the Italian word for time), it is the speed of the piece of music, represented as a number: the number of beats-per-minute, or BPM. DJs understand that the BPMs must be matched between two pieces of music if they are to transition between them smoothly during a DJ set. But there is much more to rhythm than just its speed. Given that you have an interest in DJing and dance music, it's likely you already have an instinctive understanding of rhythm and tempo, even if it's all subconscious.

First, let's look at the concept of rhythm. Try tapping a finger on something with about a second between each tap. Try to maintain an equal amount of time between each one. You feel yourself get pulled into a rhythm. The human brain will subconsciously lock onto repetitions it hears around it, particularly if the sounds are percussive (short burst sounds such as a drum or a footstep). The mind detects the pattern after a few hits and develops an expectation of when the next sound should occur. As each new tap satisfies this expectation, the brain becomes more immersed in it, creating the illusion of music. This equal-time tapping is the simplest example of rhythm. It is this feeling of expectation that, on a larger scale, allows people to time their events together and play music in groups, to clap in time at a concert or even at a sports game where there is no music to guide them.

This repetitious pulse is what we call the 'beat' of music. At its simplest form, rhythm is all about this relentless beat; one after the other in equal distance. However, the brain also gets bored by simple rhythms quickly, so at its most complex forms, rhythm is full of intricacies, subdivisions and unpredictable patterns. If you imagine each beat as the hours on a clock, the subdivisions are like the minutes and seconds that make up that hour. They are less meaningful, but still contribute to each hour. But remember, at its core, rhythm remains a slave to that strong, underlying beat.

Dance music can often be a beautiful example of this, particularly in minimal house or big-room genres. The beat will often start simple—just a **kick drum** playing on every beat, setting up that expectation for the listener. *"Ah"*, says our brain, *"good old predictable kick drum"*.

But then, another drum comes in, such as a clap or snare on every second beat, and suddenly our rhythm seems to swing back and forth between two different states; solid and grounded on the solo kick, excitement but a little tension on the clap. Soon, a high-hat loop is added to the mix, giving the rhythm its own particular character and style. It also seems to give the music more pace and energy, even though the tempo stays at the same BPM, because more rhythmic events are occurring in between the beats. But even as it grows more complex through the song, underlying it all is that simple beat, telling us where to stomp our foot or nod our head (Figure 4.10).

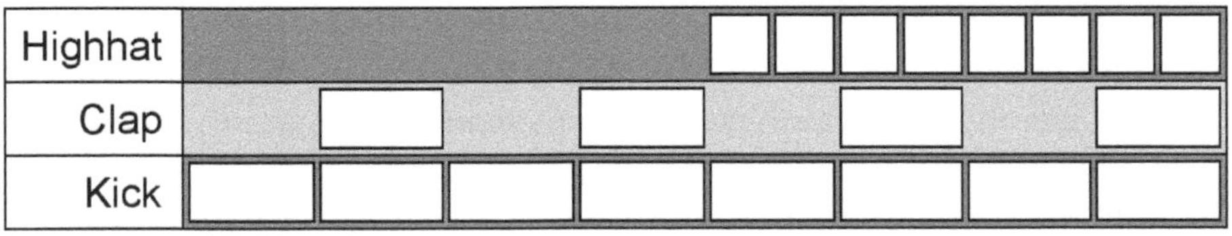

FIGURE 4.10

To look at this more technically, rhythm is essentially the pattern and rate over time around which music spreads its individual events. Similar to how scales enforce a particular pattern of notes in order to create a specific feeling, rhythm enforces a set of rules that determines where rhythmic events should fall. Are there two subdivisions between each beat, or three? Does each subdivision further divide into another two? How many of these subdivisions can be used in one beat without feeling too frantic? When events happen predictably, and inside the rules set out by rhythm, we as listeners feel the rhythm and get wrapped up inside the illusion of the music. It makes us nod our heads, tap our feet, maybe even breakdance uncontrollably.

Technically, the beats could continue to subdivide on and on forever, but there's a limit to what feels comfortable to the listener. At higher tempos of 120 BPM and up, our ears can't handle as many hits in between the beats—things begin to feel too busy and rushed. At these speeds a subdivision of four within each beat is about all we can handle. At lower tempos of 70 and below, there is lots more space between the beats to put complex patterns. Sometimes they can cram eight high-hat hits in between each beat in an R&B or trap tune—now that's subdividing!

Similar again to the idea of key and the root note, our ears hear the beat like a beacon pointing to the core of the rhythm. And like the secondary notes within a scale, the subdivisions between the beats and more complex rhythmic patterns all revolve around the beat, and determine the subtle character and feeling *within* that rhythm.

In most dance music, EDM/big-room, trance, house etc., the 'beat' is easy to identify, as there's usually a whopping big kick drum to mark each one. In other broken-beat dance styles such as drum and bass, dubstep and hip hop, it's not as simple to figure out. The 'beat' itself is consistent, but the kick drum may not always land predictably on each one. This gives these other styles a more 'shuffly' or broken-up feel, and our subconscious brain constructs the beat internally for us to follow. Thankfully, Ableton Live does a pretty reliable job of determining it for you. Also, most DJ-based music library applications will automatically determine the BPM of a track when you import your library. If you decided to use Mixed In Key, or currently use DJ playlist software such as Traktor, Rekordbox or Serato, these will analyse the tempo for you.

TIME SIGNATURES

So the human brain loves rhythm. If it likes to find patterns on the small scale between the beats, does it also like to find patterns on a larger scale? You bet it does. Our subconscious obsession with patterns means that we even like to hear beats grouped into blocks, which in music theory are known as **bars** (sometimes called *measures*).

Going back to our clock analogy: if beats are the hours on a clock, bars are like the days they are grouped into.

The brain finds it easiest to group beats into bars made up of four beats, although many different rhythm structures exist, some using very simple and even groupings, others using unusual structures that feel less natural, and therefore are harder for the brain to feel 'in rhythm' with. These rhythm structures are called **time signatures**, and determine how many beats occur per bar in a piece of music.

Most popular dance music, and indeed a lot of popular music in general, follows the most natural of time signatures: four beats per bar (Figure 4.11). In music this is called 4/4 time (four–four time).

FIGURE 4.11
4/4 Time Written in Musical Notation

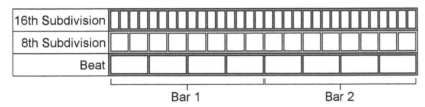

FIGURE 4.12
4/4 Time Viewed on a Timeline

The first number refers to the number of beats per bar, and the second represents the rhythmical 'size' of each beat. The number 4 represents the most common size, easily multiplied and easily subdivided; 4/4 time is simple because it explains that the music has four beats of very standard size.

You could safely say that at least 90 percent of dance music is in this regular 4-beat time signature, because it is so easy for the human brain to grasp without effort. Usually it is quite easily recognisable by a kick drum on each beat, and high-hat patterns using the rhythms that fall halfway between each beat. Each bar of music feels like it can be sliced in half evenly, and each beat feels like it can be divided evenly into two as well.

One small anomaly for you to be aware of is that music in 4/4 time can sometimes be *swung*. Swing (sometimes called shuffle) rhythm is where one of the small subdivisions will lag behind a little, rendering a looser feel to the rhythm while still obeying it. Our brains don't mind this lag, so long as it stays constant through the music. Such rhythms are very popular in genres such as jazz, rock and blues, though they typically occur on the notes between each beat in these slower styles (called 8th notes, because they make up an eighth of the whole bar). In dance music they exist on the next subdivision down (16th notes), and are commonly found in funky house music, tribal house, more quirky styles of electro and some subgenres of techno (Figure 4.12). Have a listen to Faithless—'Insomnia', 'Holding On' by Disclosure & Gregory Porter, or Nicky Romero—'Toulouse' as examples. If you import one of these tracks into Ableton Live and look closely at where the high-hat hits fall, you'll notice that every second hit falls slightly later than the grid says it ought to (Figure 4.13).

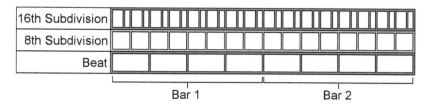

FIGURE 4.13
An Example of Swing/Shuffle Rhythm in 4/4 Time

More often than not, you can get away with merging one straight-rhythm element and one swing-rhythm element, but just use your ears to be sure. People are subconsciously a little more forgiving with interchanging swing, as they hear it all the time when the DJ mixes from one track to another. It's more important to be aware of it simply because adding straight time-based effects or busy drum elements may clash with the overall swing feel of your mashup.

Besides 4/4 time, the other time signature you will often see in other genres, and occasionally even dance music, is 6/8 time (Figure 4.14). It technically describes a bar that contains six divisions instead of four, each division being half the size of a regular '4' beat (the numbers double as you deal with smaller and smaller rhythmical sizes). What your brain in fact hears when listening to 6/8 music is two beats, divided into three smaller divisions each (Figure 4.15). It hears these three smaller subdivisions as locations for all that extra rhythmical character.

FIGURE 4.14
6/8 Time Written in Musical Notation

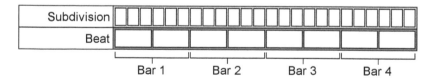

FIGURE 4.15
6/8 Time Viewed on a Timeline

Let me put it another way. If it is a genre of music that usually places a kick drum on each beat, it will still do so for a 6/8 track, placing the kick drum on the two pulses. Two repetitions of 6/8 will result in four beats, so to a certain extent we can imagine that two bars of 6/8 is the same as one bar in a 4/4 song.

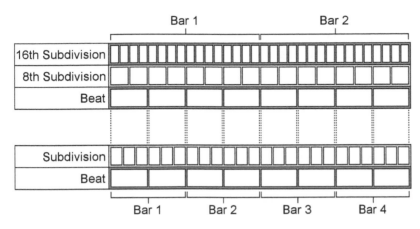

FIGURE 4.16
4/4 Compared to 6/8 Time

But each beat is further divided into three, rather than two. Therefore if a drum or instrument hit was placed halfway between two beats, it would break the rules of the rhythm, sounding clumsy and shattering the illusion of music for the listener. Because 6/8 contains an additional hit between each beat, it either requires a slower tempo or reduced sub-rhythms in order to prevent it being too busy. This generally means that 6/8 tracks and 4/4 tracks feel very rhythmically different (Figure 4.16).

This 'triplet' feeling time signature creates more of a 'marching' feel, and often exists in more underground or hard-core genres of dance music such as gabber. It also became very popular with big-room/EDM producers around 2013 with tracks such as Dimitri Vegas & Like Mike—'Wakanda' and DVBBS & Borgeous—'Tsunami'.

Being the second most common time signature in popular music, it's also very likely that you will find breakdown music you want to use in 6/8; many famous tracks are written in it, such as Katy Perry's 'I Kissed a Girl', Queen's 'We Are The Champions', John Mayer's 'Gravity', Metallica's 'Nothing Else Matters'.

What does this mean for mashups? Just that you need to be careful when trying to blend a piece of music in one time signature with another. Whilst 6/8 time has a different set of subdivisions, it will still appear to import successfully into a standard 4/4 Live session, and all the beats and bars will seem to line up on the grid. The tricky thing is that the three subdivisions inside each beat will clash with Live's grid. The most effective way to include two elements with opposing rhythms in one mashup is usually to use a 4/4 (regular rhythm) breakdown, and a 6/8 main track, or vice versa, but never allow them to actually play over each other. Instead, you pick a specific point for the mashup to switch from one rhythm to the other, and use **delays**, **reverbs**, and riser effects to smooth the transitions, and giving the listener a small gap with no subdivided hits, so there's time for their inner rhythm to reset back to a simple beat, ridding themselves of all the additional complications in between (think back to the example of tapping your finger).

Outside 4/4 and 6/8, there are many time signatures that exist in music, but as we've seen in this chapter, dance tracks are intentionally quite simple and generally unable to be written in more complicated time signatures without becoming unplayable. Let's just say that if you try and mash up the theme from 'Mission Impossible', which is in 5/4 time, you'll be in for a shock. For now, the two we covered are all you need to know.

MUSIC LOVES HANGING OUT IN BARS

As we hear music progress through each bar, we feel a sense of expectation that something should change or some event should occur as we reach it. Musicians create pleasing music by organising important events around these bars. Most of the time, they fulfil that sense of expectation by placing sounds or changes at the start of these bars in order to keep us 'in the groove' or seeing that illusion of music. These events occurring at significant points of rhythm help us keep count, feed that feeling of expectation, and give us clues as to when things will happen. For example, the instruments that play on top of the drums will usually loop within 1, 2, 4 or 8 bar patterns, in **sync** with the rhythm. Important effects or drums such as crash cymbals will hit at the start of each 8-bar grouping. A little drum fill might lead into each new eight. Perhaps at the end of each 4-bar section a guitar will do a little change before returning to its normal melody, just to keep things interesting. Another thing that dance floors expect is to hear an addition or significant musical change every eight bars, such as a new rhythmic element, vocal or synth; or for elements to drop out. See Figure 4.17 to see a representation of a standard dance music section. The blocks indicate when particular elements are playing.

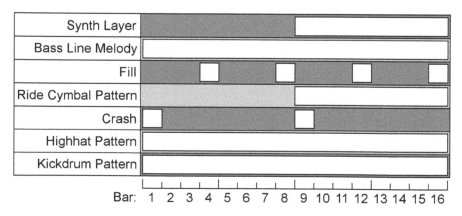

FIGURE 4.17
16 Bars of Standard Club Beats

As you may have guessed, in dance music the most important grouping of bars is eight (and to a lesser degree, four). At a BPM of 128, it takes 15 seconds to experience eight bars of 4/4, and this seems roughly to be the magic balance between being short enough not to bore the crowd, and long enough to give them the space to experience each new element or change.

As an experiment, I encourage you to put on your favourite dance track now, and see if you can figure out where the beats are, and see if you can count out the bars. Try to count out an 8-bar section.

If you find you can do it, start the track again and write down what happens every eight bars.

DJs understand that dance music works in 8-bar sections, and this is why it is, by comparison to other types of music, so predictable. Producers need to write their tracks to fit into a reasonable standard, particularly in the intro and outro, or when DJs try to mix between them they will fall out of sync and cause problems. When a DJ starts mixing a track in on the dance floor, they must start playing the new track so that its 8-bar sections coincide exactly with the 8-bar sections of the track they are mixing out of. This is so that during the time that they play together, their significant rhythmic events occur at the same moments. This allows the DJ to **fade** from one to the other smoothly, as one piece of music simply hands over the beat, bar and rhythmic duties to the next.

When making mashups, you must keep this in mind while you edit or move sections of your elements. The rhythm patterns between the elements within a mashup must be respected for the same reason as starting a DJ mix at the right bar. You must not break the illusion of the music. Trust me, your dance floor will know if you do.

You should view 6/8 music the same way. So long as you have made sure your elements work together, you can go back to treating a 6/8 piece of music based on its main beats, and think of it as 4/4. You'll notice that if you count four beats of 6/8 dance music as one bar, it still works in 8-bar sections and places its important rhythmic events in all the same spots. Another great thing about 6/8 is that Ableton Live and other BPM-detecting programs usually have no problem discerning the BPM of 6/8 music, since the primary beats themselves are in the same places.

In addition to full recordings, you will also find that vocals, riffs and any build effects that have a rhythmic element to them will not fit over the top of an element with an opposing rhythm structure. If you're in the middle of creating your mashup when you discover some pretty nasty rhythmic clashes, you may just need to zoom in and check your elements to see whether there's a triplet element in there causing chaos.

All of this tells us that to be able to blend two musical elements together, we need to make sure that when they play together, they play at the same tempo *in addition to* playing at the same pitch.

Though there are only 12 possible keys from one octave to the next, there are an infinite variation of speeds. So in order to blend elements together, you will almost always have to manipulate the speed of one to match the other.

CHOOSING THE TEMPO OF YOUR MASHUP

> When using Ableton live: the record you're using for the most part of the mashup, say the drops, keep the BPM of your mashup to this original BPM.
>
> (Dave Winnel, DJ/Producer)

Just like shifting the pitch of an element, we have the same problem when shifting its tempo up or down. In order to do it, we will need to introduce a certain amount of destruction to the part we want to manipulate.

The best way to maintain high fidelity when putting together your mashup is to set Live's session tempo to the *same BPM as the main track*. This is the one element that has had the most effort go into it, it is the most sonically dense, and it the most crucial for keeping your dance floor moving. Basically, be nice to your main track!

If your mashup will contain multiple main tracks (or 'drop' sections) that possess different BPMs, I suggest trying to set your mashup session to a BPM that also keeps all of your other elements happy.

For example, if you wanted to create a mashup that starts with one house track at 126 BPM, introduces a vocal that is 125 BPM, then transitions into a final house track at 128 BPM, set your session's tempo to 126 BPM, as you're minimising the potential audio damage to the vocal by reducing the amount it needs to stretch.

Though you could try to have the BPM shift halfway through the mashup to allow both main tracks to stay at their original tempo, this increases the risk of making mistakes in your live set when the BPM suddenly changes and pulls the rug out from under you. My suggestion is to have your BPM remain consistent if possible.

Here is how you should prioritise elements in terms of which you should keep as close as possible to their original BPM. The ones at the top need to stay as close to the original speed as possible, the bottom ones can be shifted further. Notice that it is identical to the list we saw when determining pitching priorities!

1. *Main track*—again, artificially changing the tempo of tracks with drums causes noticeable audio glitches. Try not to push your main track more than a difference of 5–6 BPM where possible.
2. *Breakdown track*—you have more room to breathe here because, once again, there are few drums.
3. *Musical riffs*—less complicated sounds mean less noticeable glitches in the audio.

4. *Vocals*—solo vocals can be pitched with an algorithm in Ableton that protects most vocal qualities well.
5. *Effects/Builds*—these contain no pitch information and can be 'resampled' instead of 'time-stretched'. We'll get into what this means in the technical chapter, but it means it can bend very far in tempo without the kind of audio glitches that damage melodic elements.

Try to think about this when you are planning your mashups, and figure out how much your elements will have to change speed. Can the element survive the change?

MASHUPS THAT CHANGE TEMPO

Didn't I just say that you want to avoid changing the tempo of your elements as much as possible, and keep your mashup BPM consistent? Yes, I did. So why would a mashup ever need to change tempo?

Whilst many of the best mashups *will* keep a consistent energy and tempo, some song combinations present themselves to you that are just too good not to use. Also important to remember is that some DJs rely heavily on slower styles of music to reach their audience, such as pop, R&B or trap. You can always begin and end your mashups at the same tempo, but without allowing some of your mashups to make a temporary tempo change in the middle, you would be forced to ignore an enormous range of music during your planning sessions. Many mashups that make use of breakdowns in these music styles will be required to pull right back in speed, down from dance tempo and into a very different feel. If done carefully, this change in speed can create a great moment by giving the audience a break from the relentless speed of club music and taking them to another place emotionally. But it has to be done right, and the other characteristics of the elements have to be an even better match than usual. You will have heard these mashups before; in fact they are produced quite a lot during time-periods where local popular music is a lot slower than club music.

For these kinds of mashups, you'll want to come up with a tempo solution ahead of time. Set a primary specific BPM just for your main sections (again, we want this to be as close as possible to the main track) and a secondary specific BPM just for your breakdown section. Most EDM-style tempo-change mashups will consist of dance sections around 125–130 BPM, with a break in the middle that slows down to a pop, rock or R&B section anywhere from 90–120 BPM. If the middle track is any higher than around 120 BPM, it might as well be sped up to match the main sections, making a single tempo mashup. If it's a lot slower than the main track, then we'll take the mashup into a dedicated slowdown moment.

Rather than allow the mashup to drop right down to the middle track's natural tempo, it's best to make an effort to push the breakdown track up a little towards the main track tempo. For instance, if a mashup drops to an R&B tune that usually sits at 90 BPM, try and push it closer to 100 BPM, so long as the audio doesn't start to glitch up too badly.

The reason you want to sit it closer to the main track speed is that there'll be less of an awkward change, and the audience will be better able to interpret it as a continuous piece of music.

We can also help hide awkward changes in tempo by using effects to cover the transitions. We'll get more into this in the technical section. Another way to help the transition happen effectively is to start moving gradually toward the new tempo in the eight bars approaching the transition.

Remember, the main thing as always is to use your ears. We talked earlier about faster songs having less dense hits occupying the subdivisions between the beats, and slower songs filling up the space between beats with more rhythm. With this in mind, if speeding up or slowing down a track sounds really nasty, it's not worth inflicting on the audience; try a different mashup combination! Try to use your ears to 'feel' what speed the slow track should come up to. Try and do so on speakers rather than headphones if you can, you'll get a better sense for it.

ONE MORE NOTE ON MAJOR/MINOR KEYS

If you have a solid understanding of keys, I'd like to revisit the subject for just a moment. I don't want you to prioritise this if you are new to the concept of major and minor keys. Once you know how to match up elements by key and get them working together, you can start looking into major and minor key combinations.

Now didn't I say you couldn't blend major and minor keys? There is one exception, and it's a very useful exception that can result in some brilliant ideas, but only if done correctly, and usually only with a little bit of luck. If we take a look at our A minor scale and C major scale for instance (Figure 4.18, 4.19).

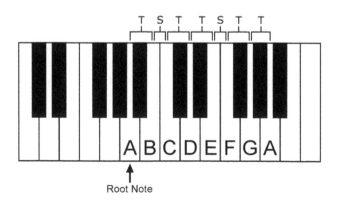

FIGURE 4.18

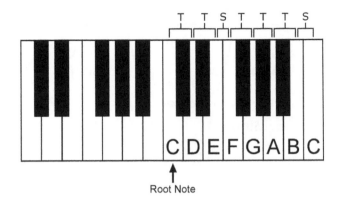

FIGURE 4.19

We can see that although they start on different root notes, all the individual notes that make up each of these two scales is identical! This partnership is called *relative major and minor keys*. What this means is that every major key has a relative minor key, and that two musical elements from each related key have the potential to coexist. This doesn't mean you can run two main tracks over the top of each other and expect it to work (because they will each be trying to communicate different root-key information to your ears), but it does mean that we can try using simple elements such as acapellas, and occasionally musical riffs, with their relative key partners. You can use the Camelot wheel (Figure 4.20) to see the relative key pairs—they share the same slice in the wheel, as well as the same number in the code.

Since we know that dance music is usually in a minor key, the most useful way to take advantage of this is to use a major vocal over a minor dance track. This is where mashups *really* get interesting, moving more into the world of remixing. If you take an acapella from a pop song in a major key, it's usually coming from an environment where it sounds more commercial, happy and upbeat. When we place this in a minor key environment, the listener's brain is being told by the rest of the musical information that the key is minor, and that the mood is cool and serious. This has a fantastic effect on the acapella. It will be fully recognisable as it follows its usual melody, but being placed in a minor environment gives it a completely new *meaning*. This effect can convert warm, happy vocals to sultry and dark; or take playful, energetic and upbeat vocals and make them feel powerful, bold and serious.

As an example, go and listen to 'Sweet Disposition' by The Temper Trap (D major), then listen to the Axwell and Dirty South remix of the same song (B minor, its relative minor key). The meaning and attitude of the vocal is completely transformed, because the music underneath commands your ear to hear the song as B minor. There are no clashes between the keys because as we've now learned, the D major and B minor scales possess all of the same

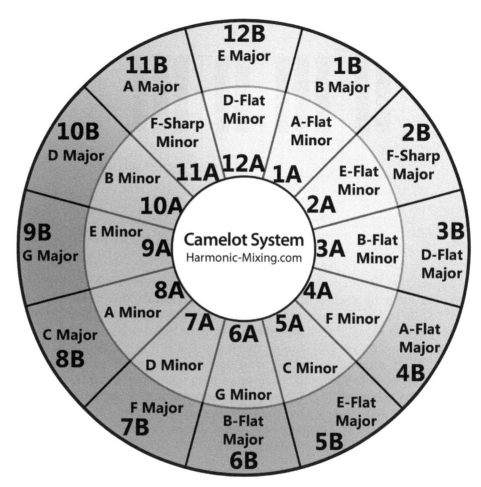

FIGURE 4.20
Mixed In Key's Harmonic Camelot Wheel

www.mixedinkey.com

notes. At no point does the vocal shatter the illusion of the B minor music underneath.

This pretty much sums up what we need to know theory-wise in order to make sure we give ourselves the best chance of composing a mashup with the least rhythm-clashing, key-clashing, audio-glitching, and audience-offending issues as possible. So, let's move into getting your equipment ready so that you can get started.

What Will You Need?

A DECENT LISTENING SPACE

First, you'll need basic acoustic treatment in your room.

Yes, most of us start out as bedroom producers and DJs, and it's not always easy to completely control your listening environment. Most of the time in mashups we are dealing with elements that have been professionally produced, mixed and mastered. This means we have a lot more leeway than if we were performing any of these individual tasks ourselves. But it is still important to try and transform our listening space into some semblance of a professional mixing environment, because we will be dealing with the balance of different sounds.

The first problem with any given room is the problem of early reflections. Sound doesn't simply exit the speaker cones and travel directly to our ears. When pressure waves are produced they fly off in every direction, bouncing off walls, ceiling, floor and objects in the room.

The loudest—and therefore—most distracting sound reflections come from the closest surfaces to you and the speakers; the walls or windows to the side and the ceiling. Imagine that the walls are a mirror, and that the reflection of the speakers is bouncing right off them. As a secondary reflection reaches your ear a split second after the direct sound from the speaker, it smears the sound information, confusing your sense of clarity and making it difficult to pinpoint the **stereo** position, tone and loudness of certain sounds (Figure 5.1).

The secondary problem is the frequency response of a room. The problem is that the box shape of most rooms creates problems in the pressure waves of sound. Due to the speed at which sound travels and the dimensions of the room you're working in, pressure waves will bounce off each wall of the room and come back, passing through each other. For some frequencies this is particularly troublesome, because the wave bounces back to its original position just as the next cycle is leaving the speaker, causing it to encourage each successive cycle. Between each opposing wall, a 'standing wave' is created, resulting in a group of bands where

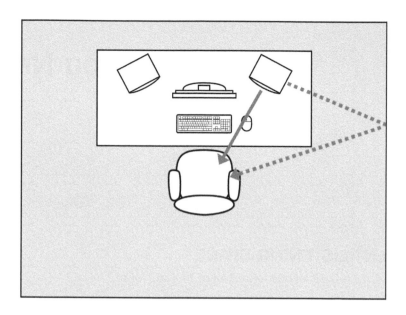

FIGURE 5.1
Direct Sound Battling with Early Reflections

pressure waves are pushing opposite each other, causing a cancellation (sound **volume** drops in this position); and bands where pressure waves coincide and combine (volume boosts in this position). These high and low pressure bands are represented by dark and light areas in Figure 5.2.

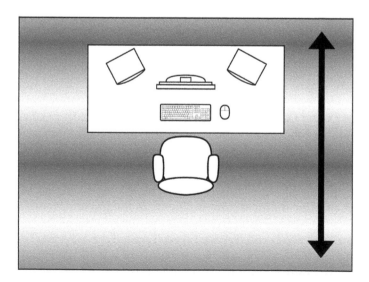

FIGURE 5.2
A Standing Wave Produced by One Single Frequency

But to make this even more difficult to prevent, each frequency of sound behaves differently. Within each frequency, pressure peaks occur at different intervals, and therefore they occur at different points between the walls.

In short, there's a whole mess of peaks and dips occurring between each parallel surface at different frequencies. Not only that, but standing waves occur between *every* opposing wall pair, combining with each other, as well as off corners (to a reduced degree). Figure 5.3 demonstrates this, and as you can see the listening position is right in a high-interference zone caused by two surface-pairs combining. These three-dimensional resonance- and dead-spots within the room are called nodes, and the frequency response you hear shifts and changes every time you move your ears through them. Even when your listening position is fixed, this makes balancing and mixing more difficult, because the different frequencies in each sound can be twisted by this frequency response and give you a false sense of each frequency's volume.

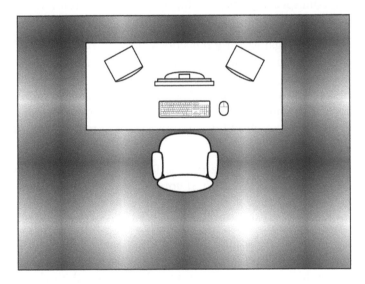

FIGURE 5.3
Room Modes Produced by One Frequency

You can spend thousands and thousands of dollars trying to battle this madness, but for mashups I suggest you just do a basic job and then use headphones to help you fill the gaps. After all, sometimes you will only have access to a laptop and headphones in the hours before a gig.

First comes speaker placement. To ensure a good balance between the left and right side of the stereo field, and a strong centre image, you want to make sure the distances between you, the left speaker and the right speaker are exactly the same. Remember, your speakers don't have to be very far away. If they're sitting on your desk, measure the distance between them. Your head and the two speakers should make a nice equilateral triangle. If it's not, move your listening position or move your speakers, even if they need to go off the desk—on speaker stands.

Also make sure the tweeters on the speaker cones are around ear height, and that the cones are pointing directly at you. Higher frequencies move in a much tighter cone as they project out from the speakers, whereas bass diffuses and spreads out in all directions, so you need to make sure the tweeters are pointing somewhere they will be heard.

Step two, make sure that if the speakers are sitting on anything, including your desk, they are not sending vibrations down into the object. Your speakers will turn your desk into a soundboard, like a guitar, and the desk will resonate with certain frequencies, causing some to ring out and taint your perception of the frequency response of the speakers even further. If you're on a low budget, even the local phone book under each **monitor** will make some dint on the vibrations, but foam specially designed for it is better. Search for 'monitor pads' or 'studio pads'.

Step three—the battle against early reflections. To set up a very basic acoustic treatment setup, look for a local DJ specialty or audio equipment store near you. If you don't have one, look online for what you need. Grab yourself around six 24x24 inch acoustic panels (or get smaller ones but more of them, to place together in the most important spots). These foam panels absorb some of the sound, mostly medium to high frequencies, and reduce how much of it comes back at you off the wall.

Some cheaper brands include Wave Panels, Foamily, as well as many smaller non-brand stores on eBay or Amazon. If you've got more money to spend and are thinking about producing in the future, consider investing in better absorption from brands such as Primacoustic, Auralex and GIK Acoustics.

Have someone help you by sliding a mirror dead against the side walls until you can see a reflection of one of the speakers. Get them to mark the spot with some tape or reusable adhesive and continue looking for the reflection of the other speaker, before moving on to the next wall/ceiling. In a regular box shaped room you should end up with two on your front wall, two on the left wall, two on the right wall, and two on the ceiling. Each pair may be close together and only require one acoustic panel to cover it.

Fasten the panels to the walls however you can. A good bet if you're renting— or even if you're not—is to use rent-safe adhesive stickers that you can find in your local hardware store. Use them to stick the lightweight panels to your walls, and there will be no damage when you take them off later. Depending on the density of the foam, you may need to PVC glue some cardboard strips onto the back of the foam panels, let them dry overnight, and then attach the stickers to the cardboard for a better adhesive bond. Also, and I know it's difficult . . . try to do the ceiling panels if you can!

Next up, the most difficult part of all—taming those bottom end frequency problems caused by the room modes. A couple of well-placed bass-traps can work magic here, but unless you are also a producer, they're a bit expensive for the level of detail you need for mashups. I'd suggest placing some thick heavy objects toward the back of the room, leaving a gap of about a foot from the wall. Perfect objects to use include thick mattresses, a bookcase full of pillows (or even just books, they absorb remarkably well) or thick curtain. If you can, place a mattress standing up in each corner behind the speakers as well. Figure 5.4 is an example of the kind of setup you're looking for, though obviously you'll need to adjust depending on where your door is.

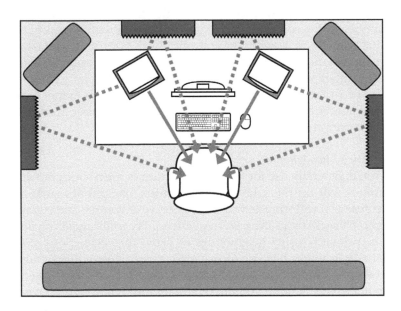

FIGURE 5.4
A Simple Acoustic Treatment Based on the Steps in This Chapter

EQUIPMENT

What equipment is required to make mashups? Here's what you need.

- *A computer*

Presumably, you already use one to access your DJ collection. You don't require huge amounts of RAM or CPU power. Most of the time Ableton will only be running a few pieces of audio at a time, with minimal effects processors working. Most laptops will be fine, so long as you have enough hard drive storage for all your music, acapellas, sessions and sample collection. Whether you use Windows or Mac is not important, Ableton Live runs on both operating systems, as do most third-party plugins you will ever need to use, and DJ playlist programs. Obviously, if possible, make your mashups on the same computer as where you keep your DJ music collection, because you will constantly need access to all of your music while creating.

- *If possible, you need a decent pair of speakers*

Sadly, the one area where you can't really go extreme low-budget is your studio **monitors** (or speakers). The desktop speakers that come with your computer are not an option! If you don't have a big budget, try to save up and invest in some entry-level studio monitors from brands such as KRK, M-Audio, Mackie, Behringer or Genelec. The reason they need to be at least studio monitor quality is to ensure you're getting a somewhat reliable frequency response and a realistic idea of what your mashups will sound like in the club. You need to be able to make choices regarding volume balances and transitions with accuracy. The telltale sign that something is wrong in your monitoring setup is when you play something out and the balance between two elements or the volume of an effect sounds totally different to how you thought it did in your listening environment. This can also be a problem with the acoustics—remember, no matter how good your speakers are, a room that does not have at least a basic treatment will produce an uneven frequency response.

- *A decent pair of headphones*

Anything you'd generally use for music production is great, but even a pair of DJ headphones will do the job. Don't use cheap headphones such as iPod earbuds because you will struggle very hard to get your volume and equalisation settings right, particularly in the bass frequencies. Try to do most of your work on speakers, primarily using headphones to listen for clicks, pops and other audio glitches you might have unknowingly created while editing audio. You can also use your headphones as a backup when problematic frequencies in your room make you doubt a volume or equalisation choice you have made.

Although I trust the sound that comes out of proper studio headphones a lot more (my favourite model is the Sennheiser HD 650), I have produced a lot of my mashups on the road with the headphones I use for DJing with very few problems. I wouldn't mix a song using them, but I'll happily make a mashup with them if I have no alternative. Just remember that in order to block out the outside sound, DJ and sound recordist headphones have closed cups, which makes it hard for the cone to move large distances, reducing their ability to reproduce low frequencies accurately. Try and check your low frequency decisions on speakers or open-back (studio-style) headphones.

- *A decent audio interface*

Though many amateur music creators stand proudly by the on-board audio that comes with their Apple MacBook, it's worth considering investing in a decent, if modest, audio interface. Though the audio that comes built in to modern computer motherboards and laptops is, admittedly, becoming more and more high-quality, you don't want it to be the weak link, particularly when it is easier and cheaper to improve than your acoustics or speakers.

The audio interface assists the computer in running the audio, as well as performing a better digital to analogue conversion than many computer sound-cards or motherboards. As a result, you get a cleaner signal, a more accurate recreation of the audio, and an output without a tainted frequency response. It is also better at handling the requirements of studio-quality headphones, which have a high electrical impedance, something computers are generally ill-equipped to handle. Unless you plan on getting into writing and producing music, a reasonably budget device should suffice. An easily accessible volume control is also very handy, if you can find a unit with one on it. I will say though, I've had to finish mashups on the road with no audio device available, and the on-board audio was enough to get the job done. If you plan to pretty much never be in your studio, you could skip this if you *really* need to. If you have put a fair bit of time into your room and have bought decent speakers, but you still notice that the frequency response between home and the club seems way off, a sub-standard digital to analogue converter may be your problem.

Jump online and have a look at what the best budget audio interfaces are in the current market. Presonus, Novation, Focusrite, M-Audio and Audient all provide reasonably good models for under $250 USD.

- *Mouse*

If you're producing on a laptop, I recommend you use a mouse, even bringing it along on the road. I've seen some producers use their touchpads, and while they have my undying respect for being able to do so, I could never make mashup creation so awkward for myself! You need to be able to work quickly and execute decisions with accuracy, and I believe having a mouse seriously outshines a touchpad in this respect. Not only in Live, but even navigating your explorer/finder **windows** and skipping through music. The fewer obstacles that sit between your brain and an action being completed on your computer, the better.

- *Ableton Live*

This one is obvious. There are a few versions of Live available, and I will go into detail in the next chapter.

CHAPTER 6

Ableton Live

To follow along with a video of this chapter, head over to 'makegreatmusicmashups.com' and watch video '1.1—Decibels, frequency and equalisation'.

Use Ableton Live. Use Mixed In Key.

(Dave Winnel, DJ/Producer)

A digital audio workstation (or **DAW** for short) is a piece of software designed to act as the creative space in which musicians can write, produce and mix music. Each DAW has a different workflow structure, different strengths and weaknesses, and different difficulties. Some are specialised, offering composers and writers the ability to notate or record their music scores in written form. Others are purely post-production workstations, and offer powerful tools for manipulating audio clips, advanced signal routing and running audio through effects processors. Most DAWs at least make an attempt to occupy the production stage in the middle, where producers record, arrange, edit and experiment with their musical ideas.

There are over 15 popular and commercially recognised digital audio work-stations in the market (over 30 if you include the less popular ones), so how do we choose which one to use? When it comes to producing full-scale music this is a very difficult question to answer. It not only requires a fair bit of testing from the user to see what works for them, but also sparks wild debates in online forums. Fortunately, for mashups our requirements are fairly specific. We require advanced and intuitive time and pitch manipulation functions, as well as the capability to run modern effects processors and detect the tempo of the audio we import.

The DAW I feel is best put together for almost this exact purpose is Ableton Live.

Ableton Live first gained popularity in the market as an application that could be used by DJs to play dance music sets in real-time, enabling the user to mix different audio streams on the fly and have them synced to a master tempo, keeping everything in rhythm. While doing this, it still allows the DJ to manipulate every track's volume and add effects in, like you could in any ordinary music production environment. It is this real-time music customisation that has allowed many artists to change their shows from regular DJ sets into a somewhat live performance. The added bonus is that people who have exactly the same requirements, only they don't need to do it live, can use it too. People such as those who want to create mashups. Remember, to create great mashups we need to take different elements, sync them to a master tempo, and manipulate and blend them in a more deliberate manner than we could with just a pair of CDJs and a mixer. In a way, mashups are merely an incredibly well-controlled DJ mix on a short timeline.

At the time of writing this book, Ableton Live is at version 9. The screenshots may look a little ancient to you if you're reading this in the year 2075 (all hail our robot overlords). If you are already familiar with Live, then you are awesome, and you should be able to skim through this section pretty quickly and get straight to the technical stuff. It is important, however, to read the sections on Warping, even if you're familiar with using Warp while playing live or creating DJ mixes. If you are new to Live, let me show you the basics so that you know which parts of the program you'll need to become familiar with to make mashups.

If you don't yet own Ableton Live, head on over to: www.ableton.com.

> A note about the different versions of Ableton Live:
>
> At the time of writing, Live comes in three forms. Live Intro, Live Standard and Live Suite.
>
> Whilst Live Suite is the preferred version, unlocking all features, making mashups is still quite achievable using Live Standard. The number of Audio Tracks, Return Tracks, Devices and other features are more limited; however, there are still quite enough to get by, and all of the Devices mentioned in this book are still included.

AUDIO TERMINOLOGY

Before we get started, let's brush up on some audio concepts and terminology that will be used in this book. The next few minutes of reading may be a little more scientific than you're used to, but bear with me. It will save you a lot of confusion later on. Let me begin with the basics of digital audio.

In the world around us, sound travels as pressure waves through air and other mediums. Digital audio is an approximation of this real-life wave-form, which is stored using *pulse-code modulation* (PCM). In recorded media, **wav** files, CD, DVD and Blu-ray all use PCM to describe the ever-changing audio signal using thousands of tiny snapshots each second. These snapshots are called *samples*. This is not to be confused with the concept of the same name where you 'sample' pieces of audio to use in other projects. A sample is basically a description of the signal's **amplitude** or **level** at a particular time. PCM is used to store or transmit a certain amount of these samples per second, at a rate called the *sample rate*. Common sample rates include 44,100 (CD quality), 48,000 (DVD quality) and 96,000 samples per second. PCM is not just used in recorded audio, but also in real-time inside a DAW as the audio is passed through the signal chain. When a sound is digitally recorded using a microphone, an analogue-to-digital converter will translate the position of the diaphragm (which fluctuates in and out according to air pressure from the sounds around it) into a sample. Later on, when each sample is translated back into an electrical amplitude at the speed of the sample rate, it causes the speaker cone to move in and out, reconstructing the original recorded sound. While reproducing sound, the speaker cone flutters between positive (pushed out) and negative (sucked in) positions. When the audio signal is silent, the cone remains at rest in the middle. This is how we record sounds, store them, and play them back as sound later.

In the digital world, each sample must be expressed as a number, which can only be stored or transferred in so many *bits* of binary information. The number of bits used to describe each sample is called *bit depth*. A 16-bit sample has 16 binary digits in it, which means it has 65,536 different possible values between its lowest and highest number. Each number describes an amplitude somewhere between the most positive signal and the furthest negative signal. For PCM signals using larger bit depths, there is even more resolution to describe each sample, at a cost of larger file size. Figure 6.1 demonstrates the digitisation of a signal with only 21 possible digital amplitudes; notice how the signal must be rounded to the nearest value.

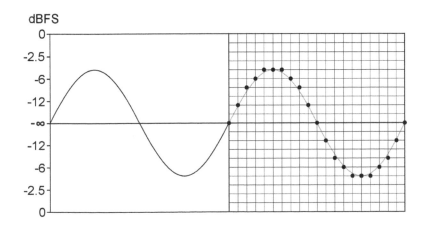

FIGURE 6.1
A Waveshape Being
Recorded to PCM Format

Decibel (dB)—The human ear is an amazing piece of sensory equipment. The power or intensity of the vibrations as they occur in the air just outside our ears can vary enormously. If you compare the smallest and quietest vibration we can detect in an ideal environment, and compare it with vibrations loud enough to make us feel physical pain, the larger vibration is around 1,000,000,000,000 times as powerful as the smaller one. This is a seriously huge difference—the loudest sounds we can handle are a trillion times more intense than the most subtle ones.

As either a physical sound pressure wave or audio signal gets progressively louder, its level keeps multiplying exponentially rather than increasing in a straight line. This is why on paper, the differences between these volumes look so immense when measured as power or intensity. However, we don't *hear* changes in volume as exponential, they sound linear to us. So in audio, we use a measurement that feels more intuitive, known as the *decibel*.

The decibel uses a logarithmic scale that helps rein in this crazy exponential effect by pulling it into a manageable range that *looks* more linear to us. It allows us to express volume using numbers that are more like the way we hear it. Another reason why it makes sense for us to use the decibel in audio is that it actually describes a *change* in volume. Technically, it describes a ratio between two numbers, and is historically used when measuring the change in anything from sound pressure level to electrical power, voltage, communications or any kind of signal. Because decibels measure a change (rather than an actual value), applying 0dB of **gain** to a signal leaves it untouched. A change of anything higher than 0dB increases its amplitude, anything lower than 0dB reduces it.

This makes it very convenient in audio because it is much easier than working in percentages or multiples of power, which, as you can imagine, becomes very cumbersome. If you imagine asking an audio engineer to multiply the kick drum by 250 percent, or to reduce that high-hat to 8 percent of its current level, it's not the most intuitive way to deal with sounds. It's much easier to say "*add a few dB here, reduce a few dB there*". And that's just for really small changes. What if you want to boost a sound in your audio program from being distant background level to being right in your face? You could be looking at as much as 40dB of boost in Live, which would otherwise be read as 100 times the voltage on an oscilloscope.

The concept of the decibel may seem a little confusing now, but you don't have to understand the mathematics of it to be able to make good use of it. The best approach is to simply develop a feel for it as you work on mashups, which you will, because you'll constantly be altering the volume of clips and tracks. Ableton Live will tell you where your volume faders are sitting in dB. Believe it or not, quite soon you'll know instinctively how many dB you want to move a sound *before* you even touch the fader. (A *fader* is the traditional name for the volume control on each Audio Track.)

A couple of things to keep in mind. The dB, when referring to signal level in a digital audio workstation, is calibrated so that when a signal is run through 6 decibels (+6dB) of amplification, the new signal will have twice the level—and twice as much movement at the speaker cone. Conversely, if you run it through −6dB of change, you halve the level of the sound. This is useful to remember because it means that if you play two full-volume sounds at the same time, you know you that the resulting signal can be up to 6dB louder at the Master Track. This gives you a clue as to whether you need to be careful of **clipping**.

The decibel is nicely calibrated for audio because the human ear can only detect a change of around 1dB in a sound—regardless of whether the sound has become louder or quieter. Interestingly, the human ear doesn't hear things in a linear fashion either. Even though +6dB doubles the signal in a DAW, it doesn't even come close to sounding like a double in perceived loudness! It's no wonder the sounds we hear can vary so much in level.

In a Digital Audio Workstation, the volume fader on each Audio Track is usually measured in dB. In Live, a blank new Audio Track begins with its fader set at 0dB, which means that the Track fader imposes no volume change on the signal before it leaves the channel. But of course, the fader can be turned down to any other level, including all the way down to complete silence, which is a change of −∞dB (minus infinity).

In the audio world, there is one other type of dB measurement you need to be aware of. What we've talked about so far is its use as a measure of the *change* in power of an audio signal. But it is also used in comparison to a reference level—in this case the maximum level of a digital signal. In this case, it is not a comparison between one signal and another, but between one signal and the maximum level—which we call *full scale*. We learned just before that in the digital domain, each PCM sample of audio can only be allocated a number between the smallest and largest number within the bit resolution of the PCM

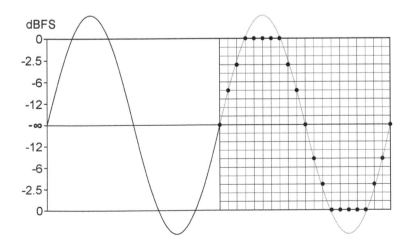

FIGURE 6.2
'Clipping'—Audio That Is Too Loud for the Limits of PCM Amplitude

format. These two numbers represent the most negative amplitude and the most positive amplitude possible either side of the zero line. In this case, if you were to describe one of your audio files as reaching a **peak** of −6dB FS, you are saying that the loudest point of the wave reaches −6dB *in comparison to the loudest possible amplitude allowable by digital audio.* Any attempt to boost an audio signal louder than 0dB FS will result in clipping, where samples straining to be recorded outside the most positive or negative amplitudes will be violently rounded back to the closest allowable value. This flattens the largest peaks in the signal (see Figure 6.2). We will talk about dB in relation to full-scale when setting up a protective limiter on the Master Track.

Frequency (Hz)—This term refers to the rate that an audio wave shape *cycles* (repeats its wave pattern). The simplest wave shape is a sine wave, which you can see in Figure 6.3. A sine wave oscillates above and below the zero line at a steady rate. The rate of this cycle, or *Hertz*, is how many cycles it performs in one second. If it takes one second to make a full cycle from the zero point to its highest point, back down past the zero line to its lowest point and back to the beginning, it has a frequency of 1Hz (1 cycle per second). If it can make the journey five times within a second, it is a 5Hz wave.

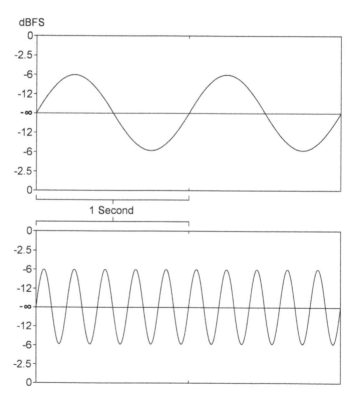

FIGURE 6.3
A 1Hz and 5Hz Sine Wave

A complex waveform is actually made up of many waves; think of any audio waveform as containing many different sine waves of differing frequencies all layered on top of one another. Some wave patterns make it easy to identify the frequencies in it, like the steady tone of a synth note or a note in a vocal performance. Other frequencies introduced during drum hits, vocal consonants or instrument overtones are harder to identify because they are fleeting or transient, but it doesn't mean we don't hear them. Even if they only have time to create one cycle, they still have an effect. Whilst the lower frequencies will look like the long, undulating hills of the wave-shape, the higher frequencies look like tiny bumps on the surface of those hills.

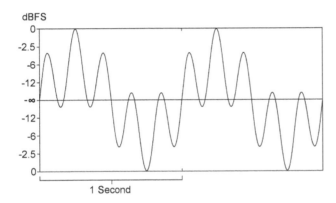

FIGURE 6.4
1Hz and 5Hz Tones Summed Together

Most audio tones we record are complicated and contain many frequencies, like the one in Figure 6.4. For instance, as a drummer hits a snare drum, the drummer uses a drum stick to hit the head, and the head vibrates causing slowly-oscillating cycles: a low frequency tone somewhere between 150–300Hz, depending on how tightly it is tuned. But at the same time, the head produces *overtones*, vibrations that occur at frequencies twice, thrice and four times the frequency of the main *fundamental frequency*. These doubling and tripling overtones change in relative volume as they go up, based on the material the head is made of, but they still contribute to the tone or *timbre* of the sound. Not only that, but a second head exists on the bottom of the snare, and the hollow chamber in between them carries vibrations back and forth, creating further resonance and therefore more sustain on the sound (depending on the dimensions of the chamber). Tightly tensed metal wires called snares are stretched along the bottom head, which also rattle as the drum is hit, adding a lot of sharp, high-speed oscillations to the sound: high frequencies. Even the drum stick will add frequencies by resonating at its own speed, if only for a split second.

Because so many sounds produce a wide range of frequencies, it means we can use effects processors to boost the volume of some, and reduce others. This is called *Equalisation* **(or EQ)**. The ear can hear frequencies between around 20Hz and 20,000Hz, interpreting the lower frequencies as 'sub, boom or rumble', and the higher frequencies as 'sizzle, hiss and sibilance'. Most frequencies we associate with the human voice reside somewhere in the middle frequencies, or 'midrange'. Using effects processors to change the relationship between frequencies will make some parts of the sound more pronounced, and others less so.

When we hear sound, the vibrations in the air are transmitted through to our inner ears, where they are picked up by tiny hair cells, each positioned to resonate and vibrate with a particular frequency. This tells our brain about the frequency content we are hearing, and enables us to locate the pitch and timbre of sound in astonishing detail.

Transient—A transient is a fast, short-duration burst of sound or audio. Found at the start of drum hits or percussive sounds, it is the initial spike in signal that in the waveform looks like a near-instantaneous jump in level. It is very important in mashup creation because you must make sure the beats of different elements are locked tightly, and looking at the transients at the start of each beat is the best way to visually line them up.

EQUALISATION

As a lot of the effects processing we must do involves using an equaliser, it's important to understand what it is, what it's used for and how it works. An equaliser is an audio processor that changes the balance between frequencies on an incoming audio signal. Each change you make to the equalisation of an audio signal is like moving a volume control, except it is for a particular area of frequencies. It is used to shape the 'tone' of a sound by altering the relationship between frequencies, boosting target areas of the incoming signal up or pulling them down. Each EQ Device will have one or more bands with a *frequency* setting that targets a particular frequency (measured in Hz), a *gain* setting that changes the volume of that frequency (measured in dB), and usually a **Q** or **resonance** setting that determines how many of the frequencies adjacent to the target frequency are also affected.

There are a couple of reasons for changing the balance of frequencies in a sound. First, if we decide to play two sounds together that sit in the same pitch and frequency range, our ears struggle to tell them apart, and are unable to hear both of them in perfect clarity. This phenomenon is called *masking*, something you would already be familiar with in real life. This is what prevents us from being able to make out somebody's voice when music is playing too loud, or what makes it difficult to distinguish the low frequencies of music while travelling in a car. Equalisation is a major part of mixing music because it is used as a compromise tool to pull back frequency information in one sound, lending more clarity to another. Without its use in mixing, we would not be able to hear lyrics with clarity or feel the most important parts of a song the way we should.

Second, it is also used to shape the tone of sounds. It can accentuate the parts of the sound the mixing engineer wants us to hear, and reduce the parts they don't want us to hear. In the case of the snare drum I explained at the start of the last chapter, you could use equalisation to reduce the sound of the head vibrating by reducing the frequencies between 150 and 300Hz, but increase the sound of the snares snapping and rattling by increasing frequencies from 7000Hz (also written as **7kHz**) and up.

Equalisation is also used to carve away problematic or unnecessary frequency bands within an audio signal. For instance, the human voice doesn't produce much meaningful sound beneath 80Hz, but microphones often pick these frequencies up anyway, recording distant traffic, air conditioner units or aircraft flying over. Even the singer inadvertently kicking the microphone stand will produce a low frequency bump on the recording. In these scenarios, engineers use EQ to **filter** away everything below 80Hz, leaving the vocal as clean as it can be.

Further to this, some electrical devices or recording equipment have a tendency to allow electrical hum into the recorded signal. Some instruments even possess unpleasant resonances that cause one frequency to spike up and distract the listener. In these cases there are very narrow EQ alterations engineers can make to pull a very narrow frequency band out of the signal.

Concerning dance music, in which the instruments are unrealistic to begin with, we have an inclination to forgive and even appreciate radical EQ movements or 'sweeps' over the whole song that would be unforgivable in live-instrument music. EQ that changes over time is commonly used this way to gradually suck the bottom end out of the music, making it feel distant and thin, before rolling away again, allowing the low frequencies to slam back in suddenly at full power.

Equalisers come in all shapes and sizes, and have different uses. Traditionally, each real-world EQ unit would only perform one or two types of EQ function, but modern plugins in the digital world such as Live's own 'EQ Eight' can perform multiple types at once.

The simplest type of EQ is known as *shelving EQ*. The user picks a frequency for the EQ to start equalising from. Then they either pick a high-shelf (to affect frequencies from the target frequency upwards) or low-shelf setting (to affect the frequencies from the target frequency downwards). Then it's just a matter of moving the gain setting to either amplify (increase) or attenuate (reduce) the volume of the designated frequency band.

The 'bass' and 'treble' knobs on hi-fi systems, a car stereo or a DJ mixer perform the very same **shelf** equalisation, they just don't tell you what the target frequencies are. Figure 6.5 demonstrates Live's EQ Eight performing shelf EQ adjustments.

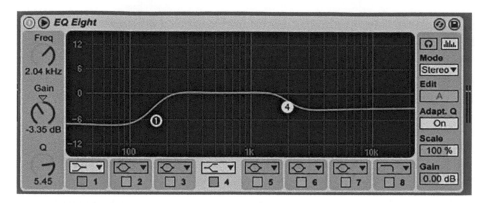

FIGURE 6.5
A High-Shelf and Low-Shelf Equalisation Using Live's 'EQ Eight'

Screenshot Appears Courtesy of Ableton

Filter EQs are similar to shelf EQs in that they target a specific frequency and affect the frequencies above or below them. The difference is that rather than performing a constant gain change across the band, they cut the frequencies out. A **high-pass** filter (or **low-cut** filter) filters out the frequencies below the target, and a **low-pass** filter (or **high-cut** filter) cuts off the ones above it. To use one, you simply choose high-pass or low-pass and select the target frequency (see Figure 6.6). Some modern models allow you to choose from a selection of slopes that determine how quickly the volume drops as you move further out from the target frequency. Usually the slope setting contains a selection of decibels-per-octave values, such as 6, 12 or 24dB per octave, though simple filters with no setting will operate at 6dB per octave. Other filters often express their slope as Q or resonance which, if turned up, adds lots of sharp boost at

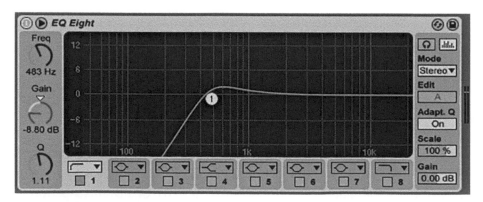

FIGURE 6.6
A High-Pass (or Low-Cut) EQ

Screenshot Appears Courtesy of Ableton

the target frequency. A slow, smooth slope is more natural to the ear, whilst a sharper slope with resonance cuts more surgically and is more noticeable. Though filter EQs are traditionally used subtly for cutting away unnecessary frequencies in the top or bottom end of a sound, they are prominently used with intentionally sharp slopes in dance music.

Parametric EQ gives the user a chance to affect a specific band of frequencies without affecting all of the high or low end information either side of it. It targets a frequency chosen by the user, and is brought up or down in level using the gain setting (see Figure 6.7). The frequencies either side of the target are affected as well, and the gain change tapers off in a bell-curve. The width of the curve depends on the Q or *resonance* setting. A high Q value tells the equaliser to make the gain change extremely narrow, causing a very sharp change at the target, a specific but fairly noticeable EQ change. On the other hand, a low Q value tells the EQ to let the gain changes expand out across the adjacent frequencies, causing more of an overall volume change, but also a very natural shape that is less conspicuous. Parametric EQ is usually the type used to boost and bring out a particular part of an instrument.

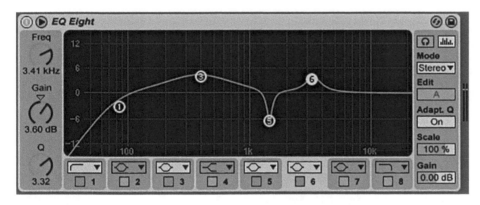

FIGURE 6.7
Parametric EQ Changes

Screenshot Appears Courtesy of Ableton

Similar to this is a *notch EQ*. Like a parametric EQ, you can use it to select a middle frequency band that you want to affect, but more like a filter, it performs a sharp cut or 'notch', which silences the target frequency. Used in mixing to remove unwanted resonances or electrical hum, it surgically eliminates a narrow frequency band from the signal. A Q setting is usually available to help manage the slopes either side of the target and avoid any noticeable side-effects.

Lastly, a *graphic EQ* is a simple Device that splits the signal into a predetermined amount of equal bands that you simply control with their appropriate fader. Whilst it is good for visually getting a handle on what your EQ is doing, it is barely used in music production due to its lack of precision. They are more useful in live music venues, tweaking the front of house EQ to account for the acoustics of the room.

Most EQ plugins or Devices allow you to make changes to more than one band at once, and usually you are able to set each band to one of the EQ types. It's not unusual for the factory default on an EQ to be pre-set with a high-pass or low-shelf band, a high-shelf band, and all the bands in between set to parametric.

GETTING FAMILIAR WITH LIVE

Let's get you up to speed in Ableton Live. Though there is a lot you can do with the application, I'm mainly going to stick to what you need to know to make mashups. I wholeheartedly encourage you to open up Live now, start a new empty session and follow along. Depending on whether you have a template set up, a new sequence may look a little different from the screenshots, but that's ok. First let's take a look at the two main views in Live—*Session* and *Arrangement* (Figure 6.8, 6.9).

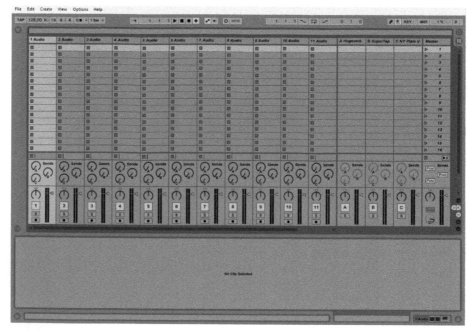

FIGURE 6.8
Session View

Screenshot Appears Courtesy of Ableton

The session screen is a summary of everything you have imported into your session, whether it is being used or not, and is great fun if you want to make up a DJ set on the spot. However, in regards to making mashups we want to create a permanent final product, and that requires us to use a timeline like a traditional DAW. So go ahead and click the 'Arrangement' button at the top-right corner of the Live window, or press the TAB key.

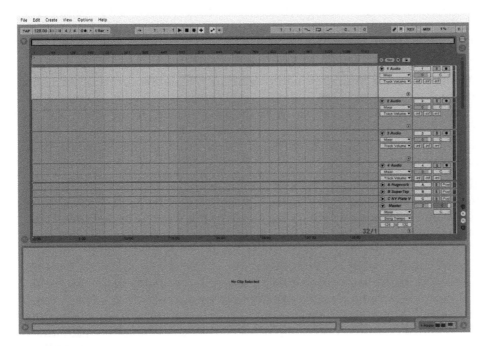

FIGURE 6.9
Arrangement View

Screenshot Appears Courtesy of Ableton

On the arrangement page we have our most important area, simply titled *arrangement*. From left to right you have your session's timeline (ranging from the beginning to the end of the session), and from top to bottom you can see all of the tracks in the session, each represented as a row.

This area is where you'll be placing all of your .wav files and .mp3s. Near the bottom you have your *Return* Tracks, where you can send audio off to be processed with delays or reverbs, and finally you have a Master Track, where the audio streams from every track come to be mixed into one final signal at the end.

Up the top of the window there are controls for *transport* (play, stop) as well as a session BPM control (beats per minute, as we have discussed in the chapter 'mashup theory'). The transport controls help you get around and control the playback of your sequence, though I highly recommend getting used to using the space bar as your play/stop button. Up the top of the arrangement window is the session ruler, where the bars of the sequence are numbered, starting at bar 1. Ableton, well aware of how much dance musicians love grouping beats and bars in even numbers, have shaded the background of the arrangement window so that it highlights groups of four. How closely you are zoomed in determines what size units are grouped into fours; bars, half-bars, beats, half-beats etc. To experiment with zooming in and out, left-click and hold the mouse on the ruler and drag the mouse up and down. To place the playback cursor, simply left-click anywhere in the arrangement window while playback is stopped, then press the space bar to start playing from that point.

On the left is the browser window, which is where you access all of Live's instruments, samples, loops, and effects processors, as well as your own sample libraries. Using the browser, you'll be dragging in ready-to-use audio such as builds, sweeps and booms to assist us in making our mashups. Try it out now— under 'categories' clock on 'Samples', and drag a sample onto an Audio Track on the timeline.

The long window along the bottom of the screen contains controls for the *Clip View* and *Device View* (see Figures 6.10 and 6.11). By single-clicking on a clip within the session, it will take the focus of this window to the current clip, or the current Devices on the track it is on—depending whether Clip or Device View is selected. By selecting Clip View, this enables you to look at the properties of each audio or **MIDI** clip and decide how you want Live to handle it within the session. For instance, if you have just dragged in a sample from the browser, you can view its audio properties and sample settings, as well as seeing a visual read-out of the waveform. Included in Clip View are controls for whether the clip is **muted** or active, speed and pitch alteration, clip volume and pan, which section of audio/MIDI is used within the clip, as well as beat markers. We'll re-visit this section in more detail when discussing *Warping*, Live's function for stretching audio, which is what we use for changing the tempo or pitch of audio.

FIGURE 6.10
Clip View

Screenshot Appears Courtesy of Ableton

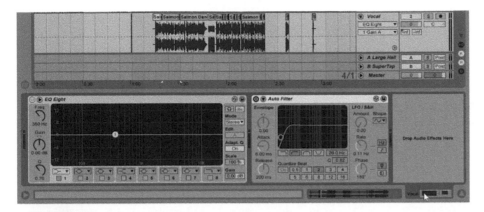

FIGURE 6.11
Device View

Screenshot Appears Courtesy of Ableton

Lastly, we can click the *Device View* button to bring up the channel effects for the currently selected track. This allows us to manipulate the audio coming through the track by sending it through effects processors such as EQ, filters, compression, limiters, delays and reverbs. Remember, anything in the *Device View* area applies to everything happening on the *whole track*, whereas *Clip View* only gives you information for *currently selected clips*.

Of course, I'd always recommend you check out the Ableton Live website to get the user manual for your current version. It's a great idea to read the chapters on Arrangement and Clip View, and to get used to navigating around the program in Arrangement view. The better you become at zipping around the software, the less interference there is between your ideas and their realisation.

For a quick run through of this section on video, check out '1.2—Getting familiar with Ableton Live' on 'makegreatmusicmashups.com'.

IMPORTING TRACKS

Considering that you will likely only be using audio, not MIDI, while making mashups, it's best to have enough Audio Tracks ready. Either go to Create (on the top menu) > Create Audio Track, or press CTRL+T (Windows)/CMD+T (Mac) while focused on the arrangement window. Then, the easiest way to import everything you want to use into your session is to drag the files directly from your explorer or finder window across to the program, and dump it onto the timeline.

Alternatively, you can create your own shortcuts to folder locations in the browser by clicking *Add Folder* under 'Places'. You can make a new shortcut for your dance music folder, your acapella folder, or your own samples. Then you can easily reach everything you need without having to leave Live.

Once you've imported all your desired elements, place them each on an individual track so that you can manipulate each element separately. Figure 6.12 shows an example of a simple mashup that consists of one main track and a vocal, with a few edits **spliced** into the elements—nothing too complicated.

FIGURE 6.12
A Simple Mashup Session with Two Elements

Screenshot Appears Courtesy of Ableton

Once an .mp3 or .wav file has been imported, Live will automatically begin analysing the waveform to try and identify a tempo for you. Wait until Live finishes analysing it before you start playing with it, or it will sometimes mute your clip, disabling the waveform image and displaying the text 'Clip Deactivated'. If this happens, select the clip and press the '0' (zero) key.

It's essential that your elements are in sync, so it always pays to double-check Live's calculations. mp3 files in particular can be a little off sometimes, as the sharp transient wave shapes on drums get a little 'smudged' when encoded as mp3, making it harder for Live to find a regular pattern. Let's look into making sure our tracks are in perfect time.

Here's your checklist for making sure your imported track is behaving itself in the session:

1. Check the start of the clip to make sure the clip begins where we want it to.
2. If the clip requires Warping, set the first Warp marker, and remove all other markers.
3. Correct the BPM if necessary.

Step 1. To check that our first marker is in the right spot, single click on the clip that represents the song you've just imported. Once you can see the Clip

View, in the sample area, zoom in to the beginning of the audio. Remember, you can do this by grabbing the Clip View ruler just above the audio and dragging the mouse down. For a dance track we want to make sure the clip begins on the first beat: usually a kick drum. Have a look at where Live has placed its *Clip Start Marker*, the small, filled-in right-hand facing arrow in Figure 6.13.

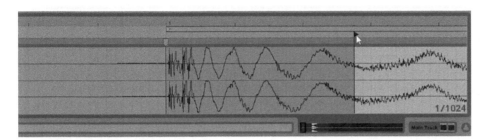

FIGURE 6.13
Screenshot Appears Courtesy of Ableton

The clip is starting just after the first beat of audio, so it needs to be corrected. It's important to check this; often Live's clip start marker will be slightly off, or in some cases, will miss the first beat entirely and have the clip begin at beat 2 of the song. For full songs that start with drums, make sure that it is on the first drum hit of the song. If it is currently placed incorrectly, move the marker to the correct spot at the start of the first beat by zooming right in to the waveform, and left-click on the audio where the beat begins (see Figure 6.14).

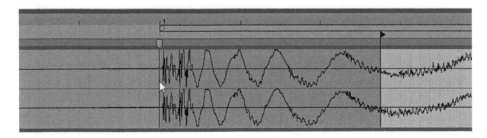

FIGURE 6.14
Screenshot Appears Courtesy of Ableton

Then right-click (CTRL-Click on Mac) and select 'Set 1.1.1 Here'. This sets the beginning of the clip to the location in the audio that you just selected, making sure that any silence or useless audio at the beginning of the wav or mp3 is not present in the session, which would otherwise shift the clip out of time and make it play late.

For other tracks that don't begin with a beat, such as effects sounds or acapellas, just set the clip start marker to the very beginning of the audio. You may have to shift the clip around in the arrangement later to make sure it is in-time, but for now, just placing the marker at the beginning will make sure that Live pays attention to the whole clip when analysing a tempo for it.

Just a note if you are currently looking at the 'main track' of your mashup—that is to say the element you have decided to base the *whole mashup's BPM off*. If you don't plan for any intentional tempo changes of your own, you can skip step 2 and 3, and instead just click the Warp button for this clip to the off position (grey). As long as you have made sure it starts on the right beat (step 1), it will already be playing at the right tempo, so it does not require tempo alteration.

If one of your other elements such as a build or acapella happens to be the same BPM as the session, you can switch off Warp after performing step 1 on these too. However, if they are an element that doesn't begin on a clear beat (for example, a vocal that begins on the fourth beat), you'll need to look at your arrangement page, place the element alongside the main track (playing at the same time) and click play just to check that it's not sitting a little in front of the beat (sounds rushed compared to the main track) or a little bit after the beat (sounds like its dragging behind the main track).

In the rare case that the element you need to correct *does* have a tempo change or tempo drifting within it, you'll need to stop here, as it will prevent the *following step* from working. You'll have to trust in Live's markers, and just keep a close eye on your mashup as you work on it.

So, if your element has a BPM that remains constant, but needs to be adjusted to sync with the mashup's tempo, move on to step 2.

Step 2. We need to use Live's time-shifting and pitch-shifting function, known as *Warping*. When you import a track into Live, it analyses the track, looking for the tempo (and any shifts in tempo, if it believes it has found any).

Make sure the analysis is correct by finding the first beat (where we placed our clip start marker in step 1). Make sure the very first *Warp marker* (a small square above the beat) is above the clip start marker. If it's not located at the clip marker, create one by zooming right into the first beat again, left-clicking on the audio to select the location, but this time right-clicking (CTRL-Click on Mac) and selecting 'Insert Warp Marker(s)'. This will create the little down-ward pointing marker, which represents a Warp marker (see Figure 6.15).

Remember, it's important to create the Warp marker right on the first transient of the beat. As I mentioned before, mp3 files produce a smudged transient and sometimes the marker will attempt to appear a few milliseconds after the actual drum hit. It's a good habit to zoom in and check. Next, we must erase any Warp

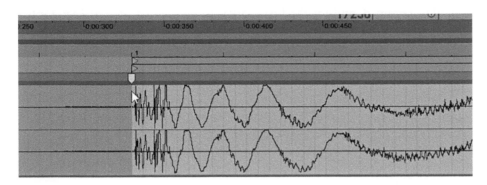

FIGURE 6.15
Creating a Warp Marker
Screenshot Appears Courtesy of Ableton

markers Live has placed *before* the one we just created. If it has, right-click (CTRL-Click on Mac) on any Warp markers before the beginning of our clip audio and click delete (these markers will be in the unused grey area of the clip—see Figure 6.16).

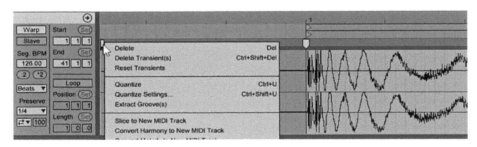

FIGURE 6.16
Screenshot Appears Courtesy of Ableton

Now that our first Warp marker is at 1.1.1 (the clip start), right-click the marker and select 'Warp From Here (Straight)'. This will erase any Warp markers after 1.1.1. Remember, we are only dealing with tracks that are a consistent tempo right now. Using the Warp From Here (Straight) function will not result in a correct BPM if there are tempo changes in the element you are dealing with.

Step 3: Now that we've made sure there is only one Warp marker on the element, we can confirm that the clip BPM is correct, so that Live knows how to treat the clip. In the sample window, under 'BPM Segment', the BPM value doesn't actually represent the speed for whole clip; rather it represents the speed from one Warp marker to the next, depending on where your selection is. This is why it's important for an element that actually has a static BPM to have *all but one* marker removed, and for the first marker to be *right on the first beat*. Only when we have one single marker can we safely count the value in the BPM box as a value that represents the tempo of the entire piece of audio.

It's very rare for club tracks made in the digital age to shift tempo, or to possess a BPM other than a whole number (for instance 125.6 BPM would be very unusual). This is not the case for earlier tunes produced on analogue gear or recordings converted from vinyl. When importing a modern club track into live, a lot of the time you'll find that Live has assigned it a BPM that reads something like 127.98 or 125.01. In this instance it's pretty clear that the BPM of the tracks are actually 128 and 125 respectively, and we can input this BPM ourselves, overwriting Live's guess.

The reason this is important is that if the clip's BPM detection is slightly off, Live will speed up or slow down the clip to match the session tempo, even if it is only by 0.01 of a beat per minute off. However, it will believe it is correcting it to a BPM value that is actually *slightly off*, meaning that it overcorrects. For instance, if a 128 BPM session uses an acapella that is actually at 128 but is being interpreted at 127.98, Live will overcorrect it to what it thinks is 128, but it will really be 128.02. The vocal, lined up to sound perfectly in time at the start of the mashup will sound out of time by the end. The result is even worse for two elements with lots of rhythm in them. By looking closely at the waveform later on in the session we would see that the drums have drifted out of sync and are no longer hitting exactly on the beats of the session timeline. That's why it is essential to make sure Live knows precisely what the original tempo is.

Sometimes you are importing an old track and have a suspicion it may drift a little bit. Prime candidates for this are old trance and house classics, or rock songs that haven't been recorded to a metronome (try mashing up The White Stripes 'Seven Nation Army'—it's a killer trying to make it stay in time!).

Try the above technique first, but if the beats toward the end of the clip come out of sync with the session's beat, undo the last step by right-clicking (CTRL-Click on Mac) the marker at 1.1.1 and clicking 'Warp From Here'. This will re-analyse the audio from scratch, but allows Live to look for tempo drifts and try to correct them by using different BPMs for different sections of the audio. If you must use 'Warp From Here', do not correct the BPM yourself. With these older tracks, don't be surprised if the BPM reads something like 126.4. Just keep an eye on the track when working on your mashup, as you may need to splice the clip in a few places (for example just before a final beats section comes back in) and manually shift the segments of the clip around to be on the beat.

USING WARP TO CHANGE THE TEMPO OF A CLIP

Now that we have Warp markers set up on every clip we intend to use in our mashup, Live will keep the clips in time with the session tempo. The great thing is that once you have your elements set up like this, you can do whatever you want with the session tempo and Live will tempo-shift everything with it—perfectly in sync with each other. If you have picked one or more of the main tracks and decided that its BPM will determine the speed of the mashup, you must now set the session's BPM to reflect it. Do this by changing the BPM up the top left corner of the arrangement page (Figure 6.17).

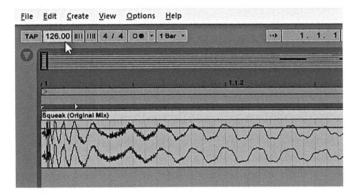

FIGURE 6.17
The Session Tempo Setting

Screenshot Appears Courtesy of Ableton

Why is changing the speed of audio so difficult? Unfortunately, because of the way audio works, if you simply get an audio program to play the audio faster or slower by compressing together or spreading out the samples that make up the waveform, it also crunches together or spreads apart the cycles within the waveform. This means that frequencies that the ear hears are raised or lowered, thereby altering the pitch as well as the speed. This causes all sorts of musical problems, as well as sounding very odd. Live's solution to this problem is *Warping* (Figure 6.18).

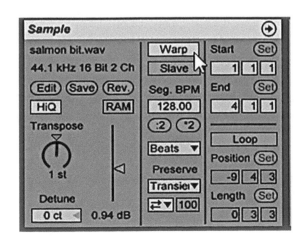

FIGURE 6.18
Enabling the Warp Function

Screenshot Appears Courtesy of Ableton

For elements that require a shift in tempo, we will have the 'Warp' button activated, but we must also choose *how* Live Warps them. Live has a selection of Warping modes that work better with different kinds of elements. Understanding them will help you get the best quality Warps possible.

'Beats' mode creates little splices throughout the clip at an interval that you specify. This essentially creates little jumps forward or backward in the audio before a rhythmically significant point in the bar or beat, and performs a short crossfade to smooth it out. Placing the crossfade just before a big obvious transient hides it so that the ear doesn't notice. When using beats mode, Live automatically sets the splicing interval to 'Transient', which tries to pick out each beat for you by looking for transients—big jumps in volume in the audio. Then it places a small splice and crossfade before the transient. This is great in theory, particularly if you are doing a live set with the software and you need to throw in elements quickly without stopping to set the beat intervals. But since we have as much time as we want, it's definitely worth setting our own interval. As we have discussed, mp3s contain awfully preserved transients, so it doesn't always detect beats the way you want it to. Often, other sounds such as bass notes, vocals and instrument stabs can set off the transient detector, resulting in badly timed splices heard as glitches in the audio. The transient setting can be very problematic for kick drums specifically, and when dealing with club tracks, the beats are simple enough that setting your own interval will yield great results—provided your Warp markers and tempo are set correctly.

'Complex Pro' mode is a more complicated algorithm that re-times the audio by attempting to stretch it evenly rather than performing quick jumps before transients. This makes it a great mode for stretching melodic parts, but is devastating to drums.

'Re-pitch' mode simply plays the audio at a faster or slower speed, without attempting to maintain its original pitch. Whilst it is the cleanest way to alter the length and speed of audio, this should only be used on elements where pitch information is not important, such as solo drums, whoosh effects, and occasionally rap (non-melodic) vocals.

Here's a guide for which Warping mode to use for which elements:

- dance tracks/main tracks during the beats sections: Beats mode, set at ¼;
- dance tracks during the breakdown sections: Complex Pro;
- smooth, long melodic elements, such as strings and pads: Complex Pro;
- short, semi-percussive melodic elements, such as synth stabs and guitar riffs: Beats mode, set to 'Transient';
- acapellas: Complex Pro (if audio glitches are very noticeable, try Beats mode, set to 'Transient');
- build effects that contain rhythms but no melody (i.e. snare rolls and timed whoosh effects): re-pitch;
- one-shots such as booms, crashes, single hits: switch Warp off; no need to re-time these;

■ one-shot reverse effects (reverse white-noise upsweeps or reverse cymbal effects); Warp off, just make sure the clips end right on the beat.

As always, use your ears when Warping audio. No matter how well selected the mode is, if you try and bring a 90 BPM rock up to 130 BPM, it's not going to sound pretty. Not only will there be audio artefacts, but it's not going to sound natural having notes and phrases that were designed to flow comfortably at 90 BPM crammed into the shorter space of 130 BPM. In instances like this, consider intentionally putting a tempo change in your mashup for the duration of the breakdown section instead.

> The record you're using for the most part of the mashup, say the Drops, keep the BPM of your mashup to this original BPM (i.e.: 128). And make sure the Warping is on Re-pitch or OFF to keep the transients sounding like the original record.
>
> (Dave Winnel, DJ/Producer)

Watch video 1.3 on the companion website, 'makegreatmusicmashups.com', to see the Warping techniques you have just read about in action.

USING WARP TO CHANGE THE PITCH OF A CLIP

Most of the time you'll have to do some subtle pitch alteration to elements to create your mashup, otherwise your choices of song combination become far too limited. In the theory chapter we talked about how far we can bend the pitch before it becomes too glitchy or unnatural; try to keep these limitations in mind when altering the pitch of a clip in Live.

First, on the arrangement page, click on the clip you want to alter the pitch of. It is important that the clip is already set to Warp and that the markers are set up. You won't be able to pitch a clip you have set to Re-pitch. In the sample box, move the 'Transpose' knob up or down, depending on where you want your clip to pitch to (Figure 6.19). The indicator just below the knob tells you how many semitones you have pitch-shifted the clip.

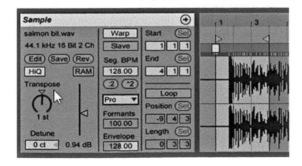

FIGURE 6.19
Transposing a Clip

Screenshot Appears
Courtesy of Ableton

To get your head around the idea of Warping to change pitch, think of 'transpose' as first speeding or slowing it down to alter its pitch (resampling), followed by the usual Warp function to return its tempo back to normal. Keep in mind that even though the Warping process is the same, the amount of Warping necessary to shift even one or two semitones is much greater. To explain why, think about how we previously discussed altering the tempo of a clip. To shift it from 125 BPM to 128 BPM, the audio will have to stretch and play +2.4 percent faster. However to pitch-shift a piece of audio up just one semitone will require the clip to play almost 6 percent faster, before Warping back to its original tempo. This shows you just how much more aggressive the Warping needs to be in order to complete a tiny pitch-shift.

Therefore, elements Warped in beats mode may not survive the usual ¼ beat interval, because the music in between them will be stretched or compacted so much that you start to notice. If for any reason you have to pitch one of your main tracks (avoid this if you can!) you will almost definitely find one beat interval isn't short enough to hide the extreme speed change of at least 6 percent. The rhythm in between each beat will be lagging behind or rushing ahead, creating a very awkward rhythm and sometimes creating double hits or *flams*. In this case, you will likely require a more extreme beat interval to combat this. Set your beat interval to ⅛. You may notice little glitches in the kick drums halfway through each beat, but this result is usually more desirable than a complete mess in the rhythm, which would break the illusion of the music and result in complete dance floor evacuation. These stretching problems are the very reason it is important to try and maintain the original pitch for your main tracks. As always, use your ears! Listen very carefully when you make a decision regarding Warping; listen to multiple parts of the element. Listen for sections with and without drums, and figure out whether you've made the right decision for the audio.

> A tip for pitch-shifting vocals in Complex Pro mode: If you find that altering the pitch of a vocal makes it sound too squeaky and 'chipmunky', or too low and hard to understand, try moving the **formant** setting around to see if you can find a better result. Live will try to shift the formant (parts of the sound that the ear recognises as human speech) back toward its natural position. It doesn't always work though, so use carefully.

EFFECTS

The arrangement view is very helpful, not only in showing you the content of each Audio Track in the arrangement, but also showing you what's happening after the audio signal moves on to the next stage, the Device section. This is a section many other programs refer to as the effects section.

If you haven't used effects before, try to imagine your Audio Track as a conveyor belt that picks up clip audio during playback, and then think of Devices as little machines you fix at certain points on the conveyor belt. Your audio signal will be fed through each Device, and emerge sounding a little bit different on the other side. The order in which you place these Devices can also determine how the final signal will sound as it reaches the end of the conveyor belt.

To view this 'conveyor belt' and see our machines at work, we must switch to the *Device View*. If the bottom panel is currently displaying Clip View, click on the *Device View* button down at the bottom-right, or use the key command 'Shift-Tab'. The Clip View area will switch to Device view. This view shows every Device currently manipulating the selected track's audio in some way, running in order from left to right. After the last Device, the audio runs through the Track Pan and Volume controls, before being sent off to the Master Track. Unless you have previously set up a template, this Device Chain will be empty to begin with. Remember—this works differently to Clip View, which is specific to one or more *selected clips*. You can have a hundred different clips on Audio Track 1, but because they are all on the same track, they will all run through exactly the same Devices. Therefore, Device View shows the Devices for the currently selected Audio or MIDI Track.

Live has a collection of Devices that come in the form of either *Instruments* or *Audio Effects*.

When using a MIDI track instead of an Audio Track in the main arrangement, an Instrument is what you would place at the beginning of the Device Chain, where it would receive the MIDI notes from the arrangement and output audio into the rest of the Device Chain. Though you can get really creative with instruments to enhance your mashups (I'll mention more on this later), for the most part you'll be dealing with Audio Tracks, which means we simply need to use Audio Effects to manipulate our signal.

Make sure your browser window (the left-hand panel) is open. If it is not, click the downwards arrow at the top-left corner of the arrangement view (Figure 6.20).

Under categories, look for *Audio Effects*. In here you will find the tools you need to help your elements work within the mashup: equalisers, filters, compressors, reverbs and delays. Once you are familiar with how to use Live's on-board effects, you can consider upgrading your effects with third-party effects, also known as *plug-ins*, as there are some fantastic signal processors out there that do a great job with specific tasks. 'Waves' plug-ins are great for most effects, however they can be an expensive option and there are plenty of great effects available as freeware. For the purpose of demonstration, I'll stick to using Live's own Devices, which are perfectly fine for getting the job done. If you own any third party VST plug-ins, you can find them under the *Plug-Ins* heading under categories. To have Live scan for your plugins, you must tell the program where they are located on your system by going to Options > Preferences > File Folder > VST Plug-in Custom Folder (Browse).

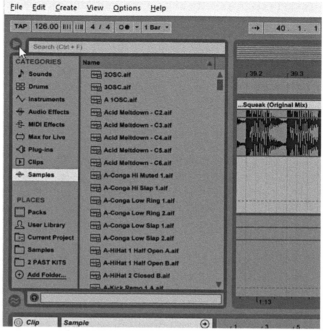

FIGURE 6.20
Opening the Library

Screenshot Appears Courtesy of Ableton

In addition to VST plug-ins, recent versions of Live make use of an additional Device source called *Max for Live.*

Essentially, it is an online world of Devices created by a community of programmers. Because they use Live's visual interface, these user-created Devices look and feel much like Live's native ones.

If you have Ableton Live Suite, access to Max for Live is included. If you own the other versions of Live (Intro or Standard), you must pay a fee for access to Max for Live. However, the Devices themselves are free. Head to the Ableton website for more information.

To place an effect into the Device Chain, simply drag it from the Audio Effects, Plug-ins or Max for Live area straight into the Device Chain. Change the order of the effects by clicking and dragging the title of an effect Device and dropping the box to the left or right of other Devices, remembering that the audio signal travels along the Device Chain (conveyor belt!) from left to right.

AUTOMATION

Though some effects can be set up and left static on the Device Chain, there will come many times when you need some of the effects parameters to move around, go up or down, switch on or off as the timeline progresses. This is particularly true for changes we need to make during transitions, such as filtering, equalisation and volume controls. To accomplish this you'll need to use automation.

To control a value over time, find the relevant Audio Track and click the Device drop-down under its title in the arrangement view (Figure 6.21).

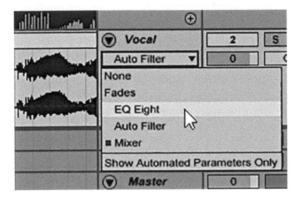

FIGURE 6.21
Opening the Automation Lane for 'EQ Eight'

Screenshot Appears Courtesy of Ableton

Select the Device you want to manipulate, such as an EQ or filter. If you want to automate a setting that isn't part of a Device it will be found under the 'Mixer' heading; controls such as volume, pan or mute. In the box below, select the value you want to control, such as gain, frequency, or send amount. An automation lane will appear over the top of the waveform on the timeline, with an automation line running through it. Double-click on the line to create a new automation point, or double-click an existing point to delete it. Now, any automation drawn on the line will move the value for you during playback. Once you have created automation for a value, you can no longer make adjustments to the knob itself in the Device panel. This is because Live can only choose to obey one source of instructions on how to handle the value. For instance, if you have created automation points on the Song Tempo automation lane, trying to change the song tempo up in the top-left corner of the screen will disable your automation. Live thinks you have decided to abandon your automation and go back to controlling the value with the mouse, and it will continue to ignore instructions from the tempo automation until you tell it to follow it again. Re-enable all accidentally ignored automation by clicking the *Re-Enable Song Automation* button up the top of the screen (Figure 6.22).

FIGURE 6.22

Re-Enabling Song Tempo Automation

Screenshot Appears Courtesy of Ableton

This is Live's way of knowing that you've finished playing around with the value and would like it to continue automating it. If you decide you don't want to have the parameter automated anymore, go into the automation lane and delete all the breakpoints. This will free up the effects parameter for on-the-fly control with the mouse once more.

To give yourself visual access to more than one automation lane per track at one time, press the plus button under the Device drop-down menu. This is really handy if you have a difficult transition to perform and want to keep an eye on both a volume and filter setting without constantly having to switch between different automation views.

While automating, it is essential that you learn to use *Draw Mode*, which enables you to draw chunks of automation that remain on one specific value for a whole measure, beat, or subdivision of beats (which length it snaps to depends on the visible grid, which is determined by the current zoom level). Hold the B key to use draw mode, or use Ctrl+B (PC)/CMD+B (Mac) to toggle it on or off. This is essential, because in dance music the timings are so robotic, so many of your automation changes will be sudden. These changes will need to occur right on the beat of the track; actions such as suddenly muting a track or pulling out all of the bass frequencies right on the point of a transition. Of course, be careful when performing sudden automation jumps, as not every Device takes kindly to it. Use your ears. It's a great habit to get into to go back and check any automation change you make; ensure that it's spot-on before moving on.

Automation can also be applied to parameters on the Master Track, where you are able to control effects on the master Device Chain, as well as global song functions such as song tempo and master volume.

That pretty much sums up the basics. The important thing is to get in as much practice using Live as you can. Navigate around, practice Warping and transposing music. Feel and hear the difference between the different Warping styles.

Watch video 1.4—'Effects and Automation' to follow along with me and see the above effects and automation lessons in action.

CHAPTER 7
Preparing for Mashups

Though organisation and preparation aren't exactly as fun as dropping a monster record or crowd-surfing, no DJ would be able to survive without them. Ten years in, I still have nightmares about turning up at a show underprepared. In clubs, you need to make sure you have all the music you need, and you need it organised so that you are flexible enough to capitalise on every idea and situation that pops up. Mashups are no different. The fewer obstacles you have between your creative ideas and your ability to make them real—the more chance you have of getting the job done properly. This is generally true when it comes to doing anything based in the world of music, whether it be writing, recording, producing or mixing. If you don't get the organisational, technical and troubleshooting tasks out of the way first, you pay for it later when you're supposed to be 'in the moment', and instead you're spending 20 minutes desperately looking for that snare build you were *positive* was somewhere in that folder marked 'cool drumz 2 use'. Now where on earth did I put that file anyway . . .

Before you embark on your mashup journey, here's what you should do to make sure you are fully prepared:

1. create a folder structure to keep things organised;
2. create a mashup key list for every recurring element you might ever want to use;
3. create and organise a samples folder;
4. collect pre-created effects and builds;
5. create an Ableton Live template;
6. label your current tracks with key information;
7. write out your first list and find your best mashup combinations.

Let's dive into these.

CREATING A FOLDER STRUCTURE TO KEEP THINGS ORGANISED

Here's what you'll need to keep things in order:

- a 'Mashup Sessions' folder on your computer. Inside this folder, create:
  - empty text file: 'Mashup Key List' (a simple text file type such as .txt or .doc);
  - empty text file: 'Current Mashup Ideas';
  - folder: 'Old';
  - folder: 'Classic Bits';
- an 'Acapellas' folder;
- a 'Finished Mashups' folder. I personally prefer to keep this with the rest of my DJ music, so that all the music that is 'ready to play out' is kept together;
  - within my mashup folder I like to create a folder for each year, just so you don't end up with a folder full of 500 mashups. (or 50,000 . . . if you *reeeally* like making mashups).

The mashup sessions folder is where *everything* you need for working on your mashups goes.

Inside this folder is where you'll save all your Ableton Live sessions, as well as keeping crucial folders and files such as your 'classic bits', current mashup ideas list and your mashup key list.

If you were writing and producing full-scale songs or albums, you would want to keep your sessions on a fast internal hard drive. The good news for you is that mashups don't require anywhere near the processing power and hard drive speed that producing music does, so I find that it's fine to put the mashup sessions folder on a secondary hard drive, if you need to save space on your main operating system drive.

The 'Old' folder (or whatever you want to name it) is useful for clearing away the session folders for mashups you have finished. Once a project is completed, simply drag that session folder into the 'old' folder so that it's out of sight (out of mind). Then you know that everything in the mashup sessions folder is active—i.e. relevant to your current job list. If you don't get into this habit, you will end up with 100 mashup folders in your sessions folder, and you will have no idea which ones are active projects—you might even end up forgetting about several half-finished projects just because you didn't see them there!

The 'classic bits' folder will fill up over time as you collect classic songs (or important pieces of those songs) that you want to use as your 'breakdown tracks'. Every time you chase down a classic song to use in a mashup, make sure you put a copy of it in this folder. This way, if you ever want to use that song in a mashup again you know exactly where to find it, and you never have to look for something twice. Often it takes longer for you to hunt down a good quality recording of a classic tune than it does to complete a mashup, so be diligent and build a collection in here.

I have suggested you keep your Acapellas folder outside of your 'mashup sessions' folder. But this is totally up to you. The reason I do this is because I also use acapellas for music production, not just mashups. I also import some into my DJ collection to play over the top of music during live sets. However, if you will only ever use acapellas for making mashups, go ahead and put your acapella folder inside your mashup sessions folder with everything else. Depending on how many you have, you may even want to sort them into folders by key.

Finally, a 'finished mashups' folder. Some people like to keep their mashups mixed in with the rest of their DJ music. If you'd like to do it that way, that's totally fine, just as long as you stick to one single method of organisation, otherwise you'll start losing track of things and some of your mashups may not end up getting imported into your gig collection! I like to create a separate folder for my finished mashups so that they're not mixed randomly with the rest of my dance music, so I know *exactly* where they'll be when I need them.

I'll explain the use of the 'Current Mashup Ideas' text file near the end of this chapter, but first let's look at the 'Mashup Key List' file.

CREATING A MASHUP KEY LIST

Wouldn't it be amazing if every time you heard a new track you could instantly picture in your mind what vocals were in key and what would work perfectly over it? Most of the time it requires great amounts of luck and memory to just have great mashup combinations pop into our head when we want them to. It will depend on all sorts of mental factors, such as what the last group of songs we heard happen to be, whether we might have recently heard an old song pop into our head or appear on the radio, or even what mood we are in. Rather than depend on sheer luck, we can greatly increase our chances of finding great mashup combinations by being organised.

Start by creating a new text file in your Mashup Sessions folder, and calling it 'Mashup Key List'. In the file, write down a list of everything you think you will ever want to use recurringly in mashups. You don't need to do this all at once— you'll be constantly remembering songs you want to use, so when you think of one just remember to write it down and add it to the file later. Include all of your acapellas, musical riffs, oldskool dance breakdowns and frequent classics. Don't worry about putting current dance tunes/main tracks in here unless they are BIG hits and you are sure that they're going to be reused. Any flavour-of-the-month stuff is just going to pollute your list. You don't want to clog up your list with things you never plan to incorporate again, and you really want to *trust* this list to give you the combinations you ask it for. List your elements out in categories based on their type (acapellas, drop vocals, riffs/breaks), and organise each category by musical key—like this example:

DJ Awesome's Mashup Key List

Acapellas:

> A-
> A#-
> B-
> C-
> C#-
> D-
> D#-
> E-
> F-
> F#-
> G-
> G#-
> X- *(Acapellas with no specific key, raps, spoken and shouted vocals)*

Drop vocals:

> A-
> A#-
> . . . *etc.*-
> X-

Riffs/breaks:

> A-
> A#-
> . . . *etc.*

Include major/minor information when recording this list. Considering most tracks are in minor keys, I usually make things easier by just writing 'A' next to an A minor track, but specify 'Amaj' for an A major track. Any acapella that doesn't possess a key—raps, spoken work etc.—just place an X next to it instead and place it at the end of the keyed acapellas, to indicate a category of acapellas that can go over anything.

Drop vocals are short little samples (usually between one beat and two bars long) that you can slice in just before a main track drops, usually a shouted vocal or small hooky phrase. The reason I keep these separate from acapellas is that you don't use them in the same way. Most of these little drops will be sampled from a full recording rather than taken from an acapella, and often not contain key information. They are much more flexible, fitting into more situations than a regular acapella or riff because of how quickly they punch in and punch out. Also, their purpose should be to amp up the crowd just before the drums kick in, similar to what a live emcee would do at a club.

Notice that I have listed all of the 'black note' keys in their *sharp* form, not their *flat* form. This is just to make organising things more straightforward, plus most modern computer file formats and music players can handle the '#' symbol in file names.

As you find more music, hear other ideas at gigs, or sporadically remember classics that would work well in mashups, remember to add them to this list. The idea is that when you're creating a mashup but you're missing one element, such as a vocal or a breakdown to go with the rest of your elements, you can browse your list and quickly find a good match. Without it, you are really depending on luck. As you build your list, you'll find you're able to create a massive batch of mashups without running out of ideas or doubling up on songs. Not only that, but it is essential that you keep a list like this if you are to become speedy and efficient at finding great combinations.

CREATING AND ORGANISING A SAMPLES FOLDER

We've discussed the idea that in addition to your main tracks, breakdowns and acapellas; you need builds and effects sounds to help guide the energy and create smoother transitions. It is important that you keep these in a simple, organised place on your computer. If you are a producer, you will already understand the value of having the right sounds in the right place, so that they're quick to access when you need them. In fact, if you are a producer you will most likely have a samples folder set up and you can simply use this while composing mashups. If you don't already have one, set up a folder somewhere on your computer. If you are using a laptop and plan to take it on the road so that you can create mashups just before a gig, I would advise you keep your samples within the computer if you have the room, rather than depending on an external hard drive.

Within your samples folder you will want organised subfolders. Organise them however they make the most sense to you. I would suggest something like this:

- Drums
  - claps
  - crashes
  - kicks
  - rides
  - snares
  - snare builds
- Effects
  - impacts
    - booms
    - sub drops
  - sweep effects
    - downsweeps and crashes
    - upsweeps
    - short reverse effects

To begin with, you may not plan to use many drums (**rides**, kicks, claps etc.). I would still strongly consider giving drums their own folder, separate from effects, because once you discover how helpful adding extra drums to build energy can be, you will wish you had created a place for them in your file system. The most important drum element will be snare builds, which you will use all the time. You may prefer to use loops rather than one-shots so that they are quick to import. Try to keep your one-shots and loops separated.

Both impact effects such as reverberated 'booms' and sweep effects such as white-noise crashes and 'upsweeps' will be used in almost every mashup where you need to transition from the main track into a breakdown. They are even more essential in mashups that change tempo. It is also useful to have a separate folder for very short reverse effects (as shown), as you need these specific effects often, and it's important to be able to tell them apart from longer reverse effects without previewing through your whole sample library to find them.

For this example I have listed two kinds of impacts: 'booms' and 'sub drops'. You'll find as you work that sometimes a big, slamming, percussive impact buried in distant reverb will work great as you drop to a breakdown, particularly in bigroom-style EDM tracks. Other times, you will find that more electronic sine-wave drops will suit your track more. Whilst most sample packs throw these into the same folder, I suggest keeping these in separate folders to make things quick and orderly.

Once you've created your file system, point Ableton Live to the directory by looking at the browser window and finding *'PLACES'*. Under this subcategory you will see *Add Folder* (see Figure 7.1). Click it, and navigate to your sample folder to add it to the list of Places you want Live to be able to view.

Similar to how you drag effects from the Audio Effects or Plug-Ins categories, you simply drag a sample to an Audio Track on the arrangement window, and it will be ready to use.

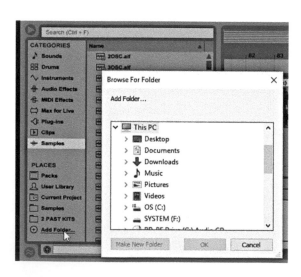

FIGURE 7.1
Adding a Folder to the Browser Pane

FINDING PRE-CREATED EFFECTS AND BUILDS

Sample packs these days are absolutely full of the kinds of effects and builds we have discussed so far.

Aware that producers are creating tracks with 8- or 16-bar build sections, many sample companies have come up with pre-programmed snare builds (sometimes called fills), or up/down-sweeps that last for a predetermined amount of bars, simply requiring you to stretch the wav file to fit the tempo of your session.

This makes it easier than ever to slip in an effects build without having to program every drum hit or meticulously time the duration of a sweep for every mashup. I'd suggest you go through a few sample packs and pick the ones you like, or the ones you think will suit your style of mashup. Collect your favourites and place them in your sample collection.

If you already produce your own music, visiting your past song sequences and exporting your own drum and effects builds is another great way to incorporate 'your sound' into the mashups. Try to keep a range of builds, some full of intense energy, some that just bubble away in the background, as one sample is not necessarily right for every mashup you do. If you deal in genres that use triplet timing, try to find some triplet drum builds too. Looking for a sample pack dedicated to one of these genres will be your best bet.

CREATING AN ABLETON LIVE TEMPLATE

> Create a folder of classic tracks that people love sorted by key. Then go through the current records you like and do the same. After that create an Ableton template specifically for making mashups. Audio channels with EQ filters, reverbs and delays already on them. Then it will be much easier to get a quick mashup going.
>
> (Ivan Gough, DJ/Producer)

Creating a Live template for mashups is a crucial step in making you an efficient mashup creator. It will also help you to avoid simple mistakes by eliminating the need to do small, annoying tasks that you would otherwise need to do every single time. The theory is simple. Basically, if you were to create ten mashups in a row, you would notice that you had to keep doing a certain set of tasks every single time without fail; creating a bunch of Audio Tracks, setting up EQs and filters on each one, setting up your master channel, setting up sends, automation etc. You need to make your mashups quickly and cut out as many technical tasks as possible. That's where a template comes in.

Begin by opening Ableton Live and starting a new session. The Live default session usually contains a single Audio Track and an Instrument Track. Delete all tracks except a single Audio Track. If there isn't an Audio Track, create one by going to Create > Insert Audio Track (CTRL+T Windows, CMD+T Mac).

Next, we want to set up that Audio Track with two Devices that we'll be using on pretty much everything; an equaliser (EQ) to pull out the bass, and a filter. Don't forget to save frequently while working on your template.

The specific EQ Device (or third party plugin) you choose to use on your tracks is extremely important. Don't use Live's 'EQ Three' Device, which creates resonances and exhibits boosts even when the gain settings are at zero. You need an EQ that is completely transparent (no change in the audio) when all of the gain values are set at 0. This is because most of the time you won't want your elements to be affected at all.

Fortunately, Live's 'EQ Eight' (which you saw briefly in the last chapter) does the job quite well. If you are already using high-quality parametric EQs for music production I suggest you use one of those, so long as it contains a 'low-shelf' dial. If using something other than Live Devices, I recommend Waves Renaissance EQ, as it's high quality and transparent. We're not using this EQ to shape sounds for tone, like we would if using them for music production, so choose transparency over character. No analogue emulations. The main duty of this EQ is to reduce the bass frequencies of the audio passing through it. Set the lowest band of your EQ to low-shelf mode, and set the gain to 0dB (neither boosting nor reducing the volume). Set the frequency to around 350Hz, with a nice broad Q of 0.7. If your EQ's Q setting doesn't go this low, just set it as low as you can. When you automate the gain of this band on your EQ to drop down in volume, all frequencies below 350 will be reduced evenly and smoothly. The low resonance ensures that the listener's ear is drawn to it as little as possible. In effect, this is the same as a DJ turning down the 'bass' knob on a mixer during a live mix.

Note: you may find that in some genres you require a slightly different frequency setting, depending on what frequency range the kick and bass sit in in that genre. Play around with the frequency setting and listen closely to figure out whether you can find the point where a gain change can seem to suck the power out of the kick and bass (Figure 7.2).

Next, we need to set up a high-pass filter (or low-cut filter). This is an effect we want to draw a little more of the listener's attention to. Rather than selecting a set frequency and leaving it there, we want to use a high-pass filter to harshly cut all the frequencies below a certain point, and then move this target frequency up or down, creating a more emphasised effect. This filter effect is widely used in dance music production, particularly toward the end of breakdowns when all the bass seems to get pulled out of the music, but we will also be depending heavily on it to transition elements in or out of the mashup.

You can use Live's 'Auto Filter' for this task, although *Xfer* have a great filter called '*DJMFilter*' that is well worth checking out. If you have played live using Pioneer's DJ Mixing Consoles, DJMFilter emulates the high-pass/low-pass filter you find on the mixer. Because filtering is very much about taste, this is one situation where it is ok to pick whichever Device sounds the coolest!

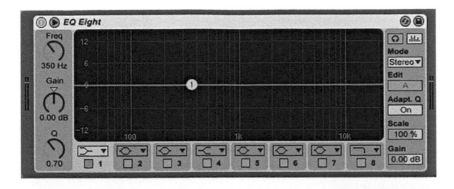

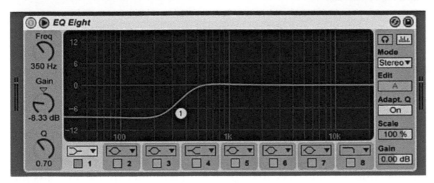

FIGURE 7.2
Setting up a Low-Shelf EQ

Screenshot Appears Courtesy of Ableton

Place your filter *after* the EQ in the Device Chain, and start with its frequency set to the lowest value. Set the filter type to high-pass, which on most Devices/plug-ins is depicted by a straight line with a roll-off on the left side. See Figure 7.3 to see what this looks like. Live's Auto Filter factory default setting is set to low-pass mode, depicted by the straight line with a roll-off on the right side, so be sure to change it. For now, set the resonance or Q to about 1, or if your filter uses dB-per-octave slopes, set it to 6dB per octave. As you become more confident with filters, you can increase the Q setting and make your filters more noticeable, but for now this setting is a reasonably safe balance between subtlety and overtness.

One more thing. Set the plugin to the *off* position by clicking the *Activator Switch*, which looks like a power button. This temporarily bypasses the Device, which is important because even a high-pass filter set to the lowest frequency (allowing all frequencies through) will have a resonance around the bottom frequency, causing unnecessary mess and loudness in the bottom end. This low frequency clutter is not often audible, particularly on headphones, but it will cause big problems when your elements are sent through the master channel, so be diligent about switching filters off when they aren't being used.

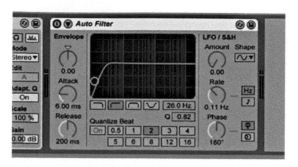

FIGURE 7.3
Setting up a High-Pass Filter

Screenshot Appears Courtesy of Ableton

> Make sure that there aren't any plugins on your channels that are screwing with your sound before going to the master. Some Filters & EQs boost or cut the highs & lows when they are in the 0% or 100% position which will totally ruin the track . . . especially in the club.
>
> (Dave Winnel, DJ/Producer)

Though we predominantly need to use high-pass filters like the one we've set up, you may find at some stage that you have a tendency to use low-pass filters too. It depends on your style and the genres of music you incorporate. The low-pass filter can be used to gradually pull the top frequencies out of the audio until only the lowest, subwoofing frequencies remain. This sounds a lot like what happens to the music as you leave a club and walk up the street, high frequency sounds becoming duller as you move farther and farther away. If you use one frequently, I suggest you place one on the Audio Track in your template. Create another filter using the method above, only this time set it to low-pass mode and set the frequency at the *highest* possible point. Again, disable the Device once it is set up; the same warning goes for low-pass filters set at the highest frequency (completely open), because there will be a sharp increase in loudness at frequencies in the upper edge of our hearing range, which are harsh and fatiguing.

Once you have your typical Audio Track set up, make sure the track volume is set at its default, 0dB. What you've just done here is provide yourself with an ideal Audio Track; one that contains all the Devices you commonly need to use so that you don't have to go setting them up every time you import an element. Now that you have this, you can duplicate the track for use with several elements. Make sure it is highlighted in the arrangement window and *Duplicate* it (CTRL+D Win, CMD+D Mac). Duplicate the track to contain as many tracks as you think you will need. I would suggest around ten to start with. If your mashups become increasingly complex as you get better at writing them, feel free to increase the number of readily available tracks in your template.

Now that we have a selection of Audio Tracks ready, it's time to set up some dedicated effects tracks. A great technique that you will use all the time to smooth transitions and enhance big moments is to send little bits of your audio out to *Delay* or *Reverb*. To do this, we need to set up a few *Return* Tracks. Return Tracks are similar to Audio Tracks, they have an audio signal path, can run Devices, and have parameters that can be automated. However, instead of receiving audio or MIDI data from clips placed on the track, they receive their audio from other Audio Tracks in the session. You can set Return Tracks up with effects such as delay and reverb, and send bits of audio to them. This is why Return Tracks are perfect for time-based effects.

Return Tracks are displayed below your regular Audio and Instrument tracks in the arrangement window. Depending on the template your new session came from, your session may already contain some, usually called Return A and Return B. Delete all Returns so that we can start afresh. Go to the menu and click Create > Insert Return Track (or CTRL+ALT+T Win, CMD+ALT+T Mac). Create three Return Tracks.

> Note: At the time of writing, users running Live Standard can only create two Return Tracks at once. In this case, I would suggest picking your most valuable two Returns from the below instructions and set up the third return's effects on a regular Audio Track, where they can be easily swapped for the effects of another Return if you need them. Try to pick your last available Audio Track so that you are unlikely to run audio through it by mistake.

Your first Return will be a *delay* effect. Delay is an effect that simply receives an audio signal and delays it by a specific amount of time before allowing it to pass out of the output. Some people call this phenomenon echo. When we send some music to a delay, it will sound like a second version of it is playing just behind the direct sound. Psycho-acoustically, it partially creates the feeling of an acoustic space by making it sound like the sound is bouncing off a surface and reflecting back to us. The shorter the delay is, the smaller the space sounds. The great thing about a delay is that we can specify a delay time that coincides rhythmically with the track, causing an effect that is not only in-rhythm, but also psycho-acoustic in nature.

With regard to your template, you will be primarily using delay as a blend effect. It can enhance a moment where a vocal finishes just before the main track kicks in, allowing the last beat of vocal to echo out over the track. We will also use it on little snippets of other elements to help transitions, but we don't want it to muddy up the bottom end when we do so. Therefore, we want a delay that focuses on midrange frequencies and has a fairly long length, or **decay** rate. If you'd like to use Live's own effect for this, I suggest *Ping Pong Delay*.

Make sure your Return Track is selected. The Device View will show an empty Device Chain, just as if it were a regular Audio Track. Drag Ping Pong Delay into the Device Chain, and set **Dry/Wet** to 100 percent wet. 'Wet' signal means the delay sounds themselves, as opposed to the original sound or 'dry' signal. It is important to set any reverb or delay on a Return Track to 100 percent wet, because if any dry signal is allowed to come back from a Return Track, it will combine with the audio coming off the original track, making the sound louder when we don't want to.

Set the **feedback** to around 50–60 percent. I usually work with tracks around 128 BPM, but if you work with tracks that are much faster, you can dial the feedback down to 40–50 percent, as the delay won't need to be as long.

Carving some bass and treble out of the delay is always good; you don't want the delay signal to create mess in the bottom end or interfere with the rhythm of the high-hats. Most acoustic spaces reverberate with reduced reflection in the low and high frequency information anyway, so it will also help the delay to sound more natural. As it'll generally be used for vocals, just a little midrange feeling is all we need. The default band-pass settings on the Ping Pong Delay are fine, just shift the frequency up to around 2 kHz. Set your delay to 'sync' to the session tempo (Ping Pong Delay is already synced by default), and set the delay interval time to one beat, or four steps. In order to respect the rhythm of the music and keep listeners in the musical illusion, it's best to sync the timing of each delay to musically significant points. Whilst you should feel free to change the delay interval to another 'in-rhythm' time on any given track, one beat is a good setting to start from in a template, as it's a good one-size-fits-all interval, and will not clash rhythmically when used over swung or triplet-feel music.

Next, we want to set up our remaining two Returns as reverb sends. Reverb (short for reverberation) occurs on every sound in existence (except for in space where there is nothing to bounce off). When a sound is created in a room, the sound pressure waves will reflect off the walls of the room continuously until they are all absorbed or too quiet for our ears to detect. When you clap your hands together in a large hall, you don't just hear your palms hitting; you hear a long trailing sound that takes a couple of seconds to disappear. The length and tonal characteristics of a particular reverberation depend on the dimensions of the room as well as the absorptive materials in it. Reverb Devices imitate this phenomenon, applying a reverb tail to recorded sounds in order to make them sound as if they are occurring in a particular acoustic space. Most reverb effects processors allow you to set a room size and a decay time to recreate the different types of real-world spaces. The decay time defines how long the reverb tail is.

The first reverb to set up will be a fairly standard 'plate' reverb, simply needed as a set-and-forget Return when have an element that is too stark, or too dry. A little reverb helps to blend it into the rest of the mashup. Be wary of the genre you usually work in when setting the style of your reverb. You'll mainly be using this on dry vocals, so have a listen to a few typical songs in your primary genre and listen to the size and lengths of the reverbs used on their vocals. Does it sound like the vocal is being performed in a medium-sized auditorium, or a giant concert hall?

As I mostly work in big-room style mashups, I chose a reverb that is most likely to blend appropriately, and sound natural in the big-room setting: a plate reverb with a reasonably large room size and a decay time of around 3–5 seconds. The idea is to have this set up in case you need to make a dry element such as an acapella feel like it belongs in the same acoustic space as the music around it. If you work in floatier, psychedelic or epic genres such as trance, you may find that an even bigger and longer reverb is more suitable, whereas a medium size room with a decay of 2 or 3 seconds will fit in with backroom house sets. Listen to styles you play a lot, and see whether you can identify how long the reverbs trail for. Close your eyes and listen closely to see whether you can *see* the size of the room in your mind's eye.

Though I use Waves 'TrueVerb' for this, there are any number of reverbs you can use. This will only be used subtly, so the Live Device 'Reverb' will do just fine. Drag in the 'Hall' **preset** from the Reverb submenu in the Audio Effect list. Alter the decay time, pre-delay and size, using your ears to compare and match a genre that is fairly standard for you. Don't forget to switch your reverb to 100 percent Wet.

Whichever reverb you choose, make sure the frequencies below around 600 Hz are cut off. You want the reverb to help an element blend into the rest of the music, but not at the cost of clarity in dance music's most important elements: kick and bass. Ensure there is no low-frequency mess by activating the reverb's cut-off. If your reverb of choice doesn't have a hi-pass (low-cut) setting, simply place an equaliser after it in the Device Chain and set a high-pass to cut all frequencies below 600. Whilst you should always be mindful of this when using reverbs, this is especially important when using them over acapellas, as a lot of the time they have noise, headphone spill, pops and clicks on them. Reverb makes these even more noticeable, as they can turn one click into a 5-second long smear across all the frequencies. Use a filter with a low Q or resonance,

FIGURE 7.4
Setting up Reverb on a Return

Screenshot Appears Courtesy of Ableton

as we don't want to draw attention to the frequency cut-off point—just subtly hide it from the listener.

For the final Return Track, you want to set it up with an absolutely massive reverb! Rather than having an element sent to it constantly, this Return will be used only at special moments to enhance transitions and help to hide edits. Repeat the process for your previous reverb Return, only this time set your room size to very large and your decay time to around 12–15 seconds. This may seem excessive, but often you need the reverb to carry audibly on through four or more bars, if you're dropping to silence for instance. If possible, use a plate reverb. Unlike our first reverb which helps elements fit together, this one intentionally conceals audio underneath it and floods the song with sound. It is particularly useful in the example just mentioned, when you want to stop the main track from playing, but still allow the *feeling* of it to spill out into a comparatively empty breakdown. Its long, lush reverb tail hides edits from the listener by reducing the intelligibility of the dry sound. Again, make sure the low frequencies around 600 and below are high-passed out, and set your reverb to 100 percent wet signal.

After you have set up your three Returns, name them (right-click the Return Tracks' title > Rename). Give them a name that's meaningful to you, like 'Delay Send', 'Plate Verb', 'Huge Verb'. Hit save now! Later on, if there are any additional delay or reverb effects that you find you often use, don't forget to add them as a new Return in your template.

Now, let's look at the Master Track. As discussed before, this is not just the final destination for the whole session's audio, but also a home for the session's global automation lanes.

On the master channel, select the 'Mixer' dropdown, followed by 'Song Tempo' (Figure 7.5).

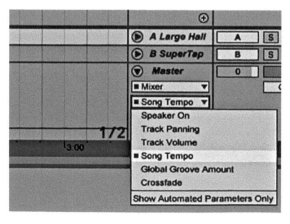

FIGURE 7.5
Song Tempo Automation

Screenshot Appears Courtesy of Ableton

As well as displaying a lane for you to automate the session tempo, you'll notice Live gives you a minimum and maximum range for this value. This is to condense your BPM range into a more convenient window, giving you better resolution to draw automation. After all, it's unlikely you'll need to automate between 1 BPM and 999 BPM in one session. Go ahead and set them to something that makes sense considering your genre. As I play mostly big-room style music, most of the club music sits somewhere around 128 BPM; and therefore my mashups will be around the same speed. I set my min and max at 125–132. You'll want to have a lower range for slower styles such as deep house and R&B, and a much higher range for trance and drum and bass. Import a few tracks from the genres you play to get an idea of the tempo range you'll need, and give yourself a few BPM either side just in case.

Up the top left of the arrangement window, there is a control for the Song Tempo. This is the control that is actually being affected by the Song Tempo automation lane. Go ahead and change this value to the approximate BPM of your primary genre. It should be a value somewhere between your minimum and maximum. Later on, when you are using this template to actually create a mashup, set your main BPM here *before* applying any automation.

Next, it's time to talk about the audio signals coming into the master channel. In mashups, most of the elements you are dealing with are mastered tracks; in other words, pieces of music that have already been pushed to their limit in terms of loudness. Every fully-mastered element you deal with will be pushing up around the 0dB FS amplitude at its loudest sections. If we consider the way digital audio works, we already know that two or more audio signals being summed (added) together will result in an even louder signal, wherein many of the samples could actually come out louder than the maximum amplitude allowed in a PCM signal. In our mashup, multiple Audio Tracks and Return Tracks are all outputting their audio signals to the master channel, meaning that when the Master Track adds them together and eventually renders them into wav or mp3 data, it will be unable to place some samples at the right amplitude and we'll get distorted or *clipped* audio.

Traditionally, 0dB FS was the absolute loudest level you were allowed to go within a DAW. Even these days, mixing or producing music where each element peaks around 0dB FS is still pretty much unforgivable! Running things this loud makes it almost impossible to avoid running into distortion problems, gain structure problems and signal processor issues long before the audio signals ever even reach the master channel. But luckily, DAWs have changed to some degree in this respect, and we don't need to be as careful while working on mashups, because we are dealing with so few elements at a time. Ableton also understands that users of Live are likely to be using fully mastered music during live performances and therefore includes a safeguard in the way it processes audio.

Aware that not everyone produces or mixes with careful gain structure in their sessions, most modern DAWs will have at least some degree of 'headroom' on every audio signal in the session. This means that while inside the session, the audio is processed with more bits, allowing for some extra amplitude beyond 0dB FS. In between each channel, each Device, and between any channel and the master channel, headroom is there to make sure the signal stays intact after signal processing or summing. This helps users out by ensuring that if they accidentally work with levels that are too loud, they won't lose samples to audio's grim reaper: clipping. The amplitude positions of the overshooting samples will be preserved all through the signal path. But there is a limit to this. Ableton Live itself is unable to record samples that peak at over 62dB FS. This is a whopping loud amount, by the way, so you'd have be pretty careless to reach volumes that loud. But you should be aware that not all third-party plugins are so generous with their internal headroom. Most EQs, compressors and other signal processors are designed to work best with levels around –15 to –20 dB FS, and when using plugins with graphical peak metres, you often can't even see what you're doing because the metres are constantly maxed out. Because standard mixing levels are so low, manufacturers don't perceive a need to drain CPU resources by including enormous bit depth to handle colossal levels. Many plugins won't even tell you if they're chopping samples off the top. If you start to run too many elements alongside each other, just be careful and use your ears to make sure none of your plugins are clipping the audio.

The other point to consider is that even if we have unclipped audio coming into the master channel above 0dB FS, at the end of the master channel Live will still need to render them into a standard PCM format. It will either render at 16-bit or 24-bit wav files where, once again, *0dB FS is the absolute maximum level*. Inevitably, it will have to round the amplitudes of every sample to the nearest 16- or 24-bit value. This means that at the end of the signal path, it's going to truncate off all those overshooting amplitudes and we're going to get some pretty nasty clipping. Ideally, we want to set up some processing on the master that activates when multiple elements are playing, and reins in those overshooting samples without compromising the sound. We also want it set up so that if a single mastered track is playing by itself, the processing will relax and allow the track through without being altered.

To do this, let's look at the Device Chain on the master channel for your template. Begin by setting up a blank 'Utility', a Device you can find under Audio Effects. Leave its settings at the default values, but click the Activator Switch on the Device to bypass it for now. Most of the time you won't need this Device. But if things get a little too loud and start cracking up when more than one element is playing, you can use the Utility to pull the entire mashup's volume back a little before it goes into the next and final Device, which is . . .

The **limiter**. All of the audio signals from every one of your tracks end up here, at the final plugin on the Master Track before the audio is outputted to your speakers, or to an exported file.

A limiter is basically a safety net to catch all of the unexpected peaks in an audio signal. If you are familiar with a compressor, it acts the same, only it doesn't allow *any* signal to pop out above a specified threshold. In mastering, that means that the resulting signal can be boosted to reach all the way to 0dB FS, making full use of the PCM's amplitude resolution and making the music as loud as other commercially available music.

Set Up a Limiter Device After the Utility on the Master Track

Different limiter plugins are designed with different uses in mind. Some start to push down the signal as it approaches 0dB, some are very transparent and only crack down once they hit zero. Some include headroom and can handle overshooting samples, others require a nice, quiet signal before they gain it up. You can use Live's 'Limiter', but you will find a lot of the time that you have to pull your utility volume back quite a bit to avoid distortion at the limiter stage. If you have a great quality limiter such as Waves L2, that's even better. It's very important that you use a limiter that will leave everything under its 'ceiling' or 'threshold' level untouched. Limiters with a 'brickwall' setting are what you're looking for. Some limiters work with advanced modes that begin to squash the audio as it approaches the ceiling, rather than actually waiting until the volume hits the ceiling first. This kind of limiter is quite musical, and very useful when mastering some types of tracks, but it's not good for mashups. It will push the dynamics around even if there is only one track playing through it, whereas we don't want to start limiting the audio until we *have to*.

If you use a brickwall limiter such as the L2, set your ceiling value to –0.2 dB, and leave the gain at 0dB. Stick with 'auto' release time, if your limiter has one, otherwise set it around 1 second.

If you're looking for a free option, rather than using Live's Limiter, try 'Limiter No.6' by Vladg Sound. It does a good job of keeping things under control without sacrificing too much volume.

   https://vladgsound.wordpress.com/downloads/

If you use Limiter No.6, make the following changes to the factory default preset to make it nice and transparent, as shown in Figure 7.6.

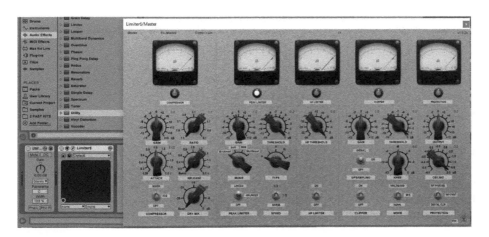

FIGURE 7.6
'Limiter No. 6' by Vladg Sound

Screenshot Appears Courtesy of Ableton and Vladg Sound

Compressor Section: Turn Compressor from 'Norm.' to 'Off' position.
Peak Limiter Section: Turn Gain to −1, and Type to 'A'.

When working on mashups later on, if you find the sound is getting far too crushed and compressed by the limiter, you can do one of two things. First, you will often find in this situation that the main track in your mashup has been mastered very loudly. In this case, your best bet is to just bring the main track down in volume a little, either by bringing all the track's volume automation points down, or by selecting all of the audio clips that represent the main track and turning the clip volume down.

If you're positive your balance is right and you'd rather not bring down the main track, your second option is to leave the session as it is, and simply bring the volume of everything down at once by activating that Utility plugin on the master channel, and bringing the gain down by 1–2dB. This way the master level is not as loud once it reaches the limiter. You'll lose a little overall volume, but it's a lot better than a distorted mashup. Remember that the limiter works by reducing the volume of level that is hitting above its threshold. Therefore, if your main track volume is fine, but you have an acapella screaming in at +5dB FS, your limiter is going to pull everything (including your main track) down 5dB when the acapella plays. This will result in a noticeable drop in energy on the dance floor. If, for some reason, you need to have your acapella that loud, you're better off pulling the main track down and giving the limiter a bit of room to breathe, so that it doesn't have to squash your audio down so noticeably.

Your template is now complete! Save it one last time. If you constantly use Live for music production, you will probably not want to make this template your default session, as you'll just end up deleting everything when you start a new project. However, if you will only be using Live for creating mashups, choosing it as the default session will add to your efficiency.

Set it up by going to Options > Preferences > File Folder > Save Current Set as Default > Save.

> If you'd like some help setting up your Live template, visit the website ('makegreatmusicmashups.com') and follow along with video 2.1—'Setting up an Ableton Live mashup template'. You can watch me create all the examples you've seen above and we can create a template together.

A final note on templates. The point of these is to make sure you never waste time setting up the same Device, same automation lane, or same Return channel every time you make a mashup. If you find that you commonly do anything that can be saved into your template, by all means add it in. This could include clips for builds or impact sounds you commonly use (put them in muted to start with), Devices for a range of interesting effects such as pitch-bending and chopping up (set to bypass mode), or even instrument tracks with sounds you commonly use to add to the atmosphere or energy of your mashups. You could keep a separate Audio Track at the end of your template as a 'storage' track—complete with disabled clips on the timeline and disabled Devices on the Device Chain.

LABEL YOUR CURRENT TRACKS WITH KEY INFORMATION

Previously, we discussed how to analyse the key of your favourite or recurring elements. Each time you embark on a new batch of mashups, you will most likely be using a new, current set of club tracks that you can't wait to play out. If you have a music playing program that allows you to create playlists, such as Pioneer's Rekordbox, then you already have a handy tool to see specific music you want, laid out in a list. You can use anything that supports playlists, even iTunes or Windows Media Player. Create a playlist for your batch of mashups. Using one of the methods from the previous chapter on finding the key, determine the key of each of the current club tracks you intend to use. Add the key to the end of the song title in the metadata (or just add it to the end of the file name if it is a wav file with no metadata). If you have found a new element that isn't a club track, such as a new pop track, feel free to include it in this list while you go, and label it with its key. This will be your playlist of everything new that you *want to include in this new batch of mashups*. Once you've found the key of each song, shift them into order from A all the way through to G#. Having them grouped together by key is important; you can compare your in-key tracks and make sure you pick the best mashup combinations. For instance, if you have three new club tracks in G minor, you can compare it to your Mashup Key List (the list you created before with all your favourite and recurring ideas), and figure out the best combinations. This way, if you only have two mashup combinations in G minor available in your Mashup Key List, you can at least see all of your current G minor tracks sitting together, and make a confident choice about which ones you'll use. It also ensures that you spread your possibilities out over your new music, ensuring you don't put the same acapella over three tracks. Otherwise, you'll only be able to play out of them in your set, and you will have wasted your time. As you come across tracks that you suspect will be long-lasting, don't forget to add them into your Mashup Key List so that you remember to consider them in the future.

WRITE OUT A LIST AND FIND YOUR BEST MASHUP COMBINATIONS

Surprise people. Combine something expected ('known') with something unexpected.

(James Ash, DJ/Producer)

Next, open up your current mashup ideas file. It should be empty at this point. Don't let this list get messy. You may sometimes still have old mashup combinations written down in there from your last batch. If you've already created all the mashups from your list last time, delete them out of the file. If you've got some sitting around that you didn't get around to, you need to make a solid decision on whether you will actually get around to making them. Otherwise, delete them. Once you've come up with your mashup 'to-do' list, you're going to burn through them—you don't want a bunch of 'maybe' ideas distracting you and wasting your time. This is just a part of being organised.

Now that your current tracks are sorted by key, you write them down in the list, including the key information. Open your Mashup Key List and begin looking for combinations.

If it's easier for you to do this with a piece of paper and a pen, go for it, but once you've decided on your combinations, write them into the text file and save it so that they don't get lost—and so that they go with you wherever your computer does.

Next comes the fun part: deciding which of your favourite mashup elements will work with your list of new music. Use the guidelines from before to check whether the elements truly belong together, based on energy, genre, emotion, and appropriate time of play. These decisions are creative ones, and depend very much on your DJ style. For some DJs, the most important factor is to use the newest, most current tracks. For others, it is totally acceptable to play older music as their main tracks, so long as they are stronger songs with better moments. Some DJ/producers will try to limit themselves to using only music that successfully supports their musical styling, or work with the acapellas and musical moments from their releases. All of these decisions must be taken into account when picking new mashups, because inevitably, each time you decide to match a recurring mashup element, such as a classic breakdown or acapella with a new track, you will likely be removing an older mashup that uses it from your DJ set. You need to make sure the newer mashup idea is worth replacing the old one, or you'll find you just keep playing the old one in your sets anyway, and have wasted your time. It's only worth making the mashup if you're going to reach for it during that moment of decision in your set.

If one idea doesn't grab you straight away, write down multiple potential combinations for each current track, and listen closely to each option using your music player. Play a snippet of your break track or acapella, then play a snippet of your main track. Try to imagine what the mashup will sound like, and ask yourself whether the elements belong together, which combination feels the best? This, by the way, is a great time to learn the keyboard shortcuts of your music player! Being able to zip around between songs and skip straight to the peak sections helps you experience what it will be like to hear one song transition to another. Pay attention to your emotions and excitement levels while you do this—this is where your inspiration will come from.

While you pick your combinations, think about the moments you want to create. As well as writing down combinations, write down any ideas you have on how the mashup should be arranged. If at any point you think *"How cool would it be if this bit went into this bit . . . etc."*, write it down! The moments you will create are even more important than the combinations. You're going to listen to so much music before you're finished with your list that you'll find it hard to remember if you don't write it down.

Here's an example of something I'd write: *'Have [main track] filter up just before the break down, use long reverb, wait 2 bars, drop into [break track]. At end of breakdown use vocal with some pitch bend effects.'* Though experimenting with a Live session containing all your elements is great for finding ideas you didn't expect, often the first 'big moment' idea that leaps into your imagination is the best one.

Once your mashup to-do list is completed, it's time to get serious, and work through them one by one.

CHAPTER 8

Making Mashups—Part 1

Arrangement and Sourcing Audio

Now that we have covered the preparations, it's time to drag in some music and start creating. Before we begin, remember that above all else, the initial concept is more important than the techniques you use to make them reality. Try to find great combinations of music, and you won't have to do as much work to make them fit together.

This chapter is all about teaching you the concepts and individual techniques that you can use to take your mashups to the next level.

First, we will cover a few points on arrangement, then get into individual techniques useful for making things work together and creating strong moments.

ARRANGEMENT

> Nothing matters more—It's gotta work on the dance floor.
>
> (Phil Ross, Commercial DJ)

Arrangement is the art of presenting a musical idea. Great songwriters, composers, bands and producers have long understood that you can have a brilliant melody, an amazing set of drums and instruments, and a mind-blowing vocal performance, but if they are not arranged together in a way that makes sense, it doesn't work. Music, in particular, has to tell a story. It must follow an arc that presents the elements in their best light, and it has to strike a balance between exposing the listener to each idea and keeping things interesting. The same goes for mashups; you can have two or more pieces of music that absolutely belong together, but if you put them together in the wrong order or ruin the best bits of each, you aren't doing each element any service. Instead, you will be standing behind the DJ booth scratching your head, wondering why the dance floor is deserted.

Thankfully, the songs we use in mashups already contain the instruments, drums and vocals painstakingly planned out and arranged. So long as we try to preserve the way that the songs want to unfurl themselves, they will do the work of guiding the listener for us. All we have to worry about is what order to place our tracks, and when to transition from one to another.

Pick the first project on your mashup to-do list, open up a new Ableton Live session with your mashup template, and import all your elements, following the steps in the chapter on Live Basics.

Once everything is imported and ready to go, listen through each element and decide how you want to approach your arrangement.

Think clearly about this, ask yourself:

- How do I want my mashup to begin?
- How do I want it to end?
- What big moments can I include here, and how do I build up to them?
- How long should the whole piece be?
- How many breakdowns do I want in this mashup?
- How can I prevent the listener from getting bored?

If, while creating your to-do list, you have already come up with some great ideas for how the music should be arranged, then follow your initial inspiration and simply start putting it together. If not, spend a couple of minutes experimenting!

Experimenting is one of the most enjoyable parts of writing mashups. It teaches you a lot about the potential of what they can do. Try dragging your vocal over a section you didn't expect it to work over. Try using a part of the vocal that doesn't seem important. Try cutting up and changing the arrangement of the main track so that it doesn't do what the audience expects. It is important to have a little 'play' with your elements before getting to work, to settle in to them. It will help make sure you've found the best potential moments you can. Sometimes the most unexpected ideas announce themselves by accident, giving you the opportunity to present something to your crowd in a way they've never heard before.

If you are using only partial sections of a song, such as a small section of vocal from an acapella, make sure you pick *the best bit*. Take the time to listen through your acapella and pick the most appropriate verse or **chorus**. If you're using a classic track that contains two breakdowns but you only intend to use one, pick the best one—or pick half of one and half of the other; whatever best serves the needs of the mashup. Keep in mind that when using a verse from a song, the first verse is generally the best one to choose because people associate this with the start of the song, and are likely to know it better than a second or third verse. For many classic acapellas, choruses near the beginning are sung in a solo voice, whilst choruses at the end are performed with layers of background vocals and ad-lib sections. Figure out which sections are most appropriate in the context of your mashup.

For more ideas on arrangement, find some sets from your favourite DJs, particularly ones who bring great mashups into their sets. Listen closely to which moments work the best, why they work, and what was necessary to build up to them. You will start to pick up ideas, such as how long you need to let one element play for before another is introduced, or how DJs have used editing to extend or shorten their breakdowns, depending on what kind of audience they are playing to or what setting the set is performed in. You'll get a feel for how far the DJ is willing to pitch-bend a particular element, or how much tempo change they feel comfortable with.

The biggest thing is making sure the mashup doesn't get boring.

(Ivan Gough, DJ/Producer)

LAYING ELEMENTS TOGETHER

To make any pair of elements feel like they belong together, we need to think back to what we learned in the chapter about theory. First, you must decide what tempo and key your mashup is in. As we've learned, the main track(s) carries the most weight in these two decisions. Next, if your elements require any tempo or pitch alteration to become compatible, that's the second step. Then we know everything is key-compatible and that we're able to see our musical elements in terms of beats, bars and 8-bar sections.

It's crucial at this point to get to know your elements. Forget about laying them together for a moment and give them their own space on the timeline, or press the solo button on the Audio Track to listen to one without the other. I like to put my main track at the start of the timeline since the mashup will eventually start with that anyway, and place my break element some distance after it (leave them on their individual Audio Tracks though). Now you can listen to each element on its own.

Analyse—what arrangement does each element conform to? With time, you'll learn to read it just from looking at the waveforms, but it's good to listen through anyway, as it will spark ideas. Low end frequencies are represented as big movement in the waveform. So, for dance tracks, usually the thickest/darkest sections are the loudest and have drums or bass-line in them, indicating 'drop sections'. Breakdowns will look like thinner, lower level sections in the middle with less transient bursts, whilst the thinner sections at the beginning and end are drum intros/outros with modest bass presence.

It's wise to listen through each element and determine for yourself which sections are which. It's also a great time to perform some splits in the audio to help you see where each section ends. Create an edit in a clip by clicking on the appropriate point of the clip and *splicing* (CTRL+E Win, CMD+E Mac).

Figure 8.1 shows a main track that has been split into *intro, breaks, drops and outro*. Notice how it will make it easy to see which sections to mute or move around. Once you have better understanding of the arrangement of each element you will be able to form a plan of attack for the mashup. The break sections from your breakdown element should now be conveniently split at

FIGURE 8.1
A Main Track Split According to Sections
Screenshot Appears Courtesy of Ableton

the section points, so you can simply drag them over to line up with the comparative sections in the main track.

When learning about rhythm, we discovered that it's incredibly important to preserve the rhythmic patterns within each piece by lining them up to coincide as much as possible, to keep the 'illusion' alive. Think of your elements as actually having very few options in which they can successfully be placed. Line up your 8-bar sections as well as you can, particularly during the bigger, dancing sections. Each time a new element comes in, the listener should not discover that crash cymbals, booms or drum-rolls are hitting where they shouldn't, or that changes in the music are occurring at odd places. You should easily be able to switch from one to the other and be hearing the same beats, bars and bar groupings continue even while the music changes. Keep that illusion alive, baby!

Figure 8.2 shows a fairly simple mashup from the beginning up to the second 'drop'. Here's a little insight into how it's put together. The two songs used are both dance tracks, so even the breakdown element, which is a remix of a commercial song, is also set up like a dance record. In this case, it's the breakdowns I wanted to use, but the bonus is that it's already arranged like a dance track, making the alignment much easier.

To show you what I mean, I've treated the song as a series of mini-sections, and labelled each with a number. In this example, each one is exactly eight bars long.

1. Section one is simply eight bars of intro of the main track.
2. The breakdown track starts to filter and fade in. I'm using eight bars of its breakdown, so all the rhythm markers align. As the section comes to an end the main track quickly filters out.
3. As I stop playing the main track, the breakdown track has been arranged so that a vocal comes in. It doesn't seem sudden, because I've already been leading up to it in section 2. It feels like a regular breakdown to the audience, who at this point probably have no idea it's a mashup.
4. I loop a piece of the breakdown so that it sounds like the final note of the melody is being held for another eight bars. Underneath, the build section

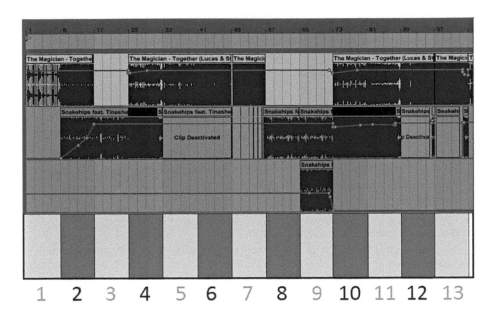

FIGURE 8.2
A Simple Mashup Arrangement

Screenshot Appears Courtesy of Ableton

of the main track comes in, hyping up the audience. At the end of this section, I allow the most final vocal phrase of the chorus to play for one bar.

5. I cut out the breakdown track, and the main track (which is kicking in with beats) plays at full volume on its own.

6. The main track follows its natural arrangement and continues for another eight bars. In the last two beats, I drop it out for effect and play a little snippet of the breakdown track vocal, just to remind people of the breakdown, and make it feel more cohesive.

7. Another eight bars of main track, with more frequent snippets of the breakdown over the top to hint to the listener that the break song is coming back in a moment.

8. As promised, we return to the breakdown track. To smooth the sound of the main track stopping, it automates to a big reverb and a delay on the last beat of section 7, so that its sonic flavour echoes into the breakdown, hiding the edit and making it feel like one piece of music. I've been hinting toward the vocal so that I can bring it in straight away. Though most of the time it's better to have a little break from the vocal, the mashup was for quite a commercial club and so I didn't want to risk going too long without it.

9. The breakdown track actually *wants* to go into a build section there, but I want my mashup break to go longer, to be different from the first. So I repeat the same eight bars, but on another track I use a duplicate of eight bars of its outro, which is mostly drums, because I need a rhythm pick-up.

I use the outro drums from the breakdown track because I know they'll mix rhythmically and stylistically over the breakdown section. I use a pick-up to prevent the audience from getting bored of the exact 8-bar repeat.

10. I go for the looping vocal idea again, only this time I extend it to be twice as long to create more anticipation before the next drop. Luckily, the producers who put together the main track produced a build that goes for 16 bars as well, so I can just use that rather than having to add elements of my own.

11. The build goes into its second eight. The breakdown track continues looping the vocal but a pitch-bending effect starts creating a rising feeling in the loop so that it doesn't become boring. At the end, I switch off the pitch bending Device and it is once again allowed to play the most important vocal phrase in full.

12. The anticipation is released as the main track kicks in. Some of the vocal phrase repeats at the end of the eight.

13. Another eight bars of beats. From here it naturally progresses into the main track's outro.

This may seem complicated when written, but believe it or not, this is quite a simple idea. It pretty much makes use of the arrangement that was already there in the main track, which the producers have put their hard work into making sure progresses properly. The mashup really just replaces its breakdowns with another song, and uses some simple hinting and building techniques to guide the listener through the changes and accept them as one musical piece. It is easy to stay true to the 8-bar sections proposed by each element, because they are both dance tracks. There's very little risk of the dance floor getting confused because of the even blocks of eight.

As you progress, you'll discover that there are ways to avoid operating strictly in 8-bar blocks, but it's something you need to test out carefully with your audiences. The more non-mathematical your mini-sections are, the greater the chance of pushing them away. If you want to sneak in the occasional 4-bar section of instrumental breakdown before the vocal comes in, that has a better chance of working than shortening one of the 8-bar sections of dancing into six bars, for example. Sometimes it can be effective to drop to silence as well, which we'll cover later in the chapter. A silence between a drop and a break is a good time to cut out bars because the listener's internal timer is much more inclined to reset without resisting.

Figure 8.3 shows an example of quite a complicated mashup (in terms of rhythm, anyway). Notice that many of the sections are uneven. So much more work had to be done to make things feel right, using techniques described in this chapter. There is a lot more filtering, chopping and help from rhythm loops in this one.

1. & 2. Both are 8-bar sections of main track intro.

3. The vocal for this commercial track comes in just before the start of an 8-bar section, so if section 4 came directly after section 2, the first few words would be blocked out by the intro, which is loud and busy. So I had to

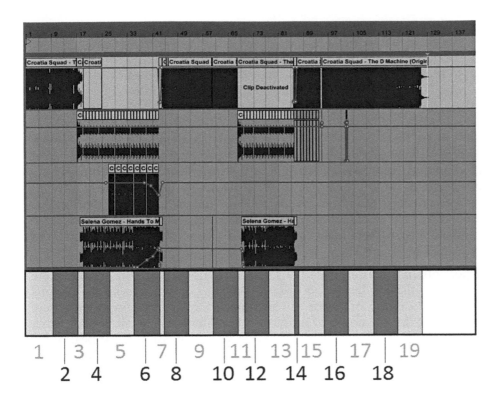

FIGURE 8.3
A More Complex Arrangement

Screenshot Appears Courtesy of Ableton

include two bars here to allow the drum to reverb out into the break, and give the audience a little bit of instrumental from the break track.

4. This is where the vocal section technically starts, though a few lyrics have already been sung. The break was very minimalist, creating too much of an energy drop, so I used a little slice of the main track outro to give it a little more drum energy.

5. The vocal continues, but the energy doesn't increase enough, so I add another loop from the main track, which is from a big loud section, but the bass is filtered out so that it doesn't kill the 'breakdown' feeling.

6. The elements all high-pass filter up a little to create anticipation. After all, the breakdown track wasn't supposed to feel like a dance track, so it needs a little help. It's timed out so that the most crucial 8-bar section of vocal plays through here, to give it the best chance of building effectively.

7. Though the mashup could have just kicked into section 8, it really needed the main line of the song dropped in beforehand. It's so important to the mashup that it surprisingly feels totally fine adding the extra bar. This is one of those times where vocals are so important they can bend the rules a little.

8.–10. Drop section. Off screen, there's actually lots of little vocal chops adding to the rhythm in 9 and 10.

11.–17. Pretty much a repeat of sections 3–10, only the break is eight bars shorter to keep it exciting, and the second breakdown chorus was exciting enough that it didn't need that second loop from the main track.

18. & 19. The main track carries on into the outro the way the producer intended.

You can see that when you're combining a dance element with a non-dance element such as pop or rock, you need to work extra hard because the non-dance track doesn't automatically follow the same format. You will often have to repeat sections, choose which breaks will actually appear in the mashup or find ways to make the break feel like it's 'building up' to a dance section. You also have to sacrifice big sections of the song to fit it into those small breakdown zones, so you'll need to make decisions on which sections are most important to the audience.

Importantly, when using an acapella, don't assume the first lyric is sung on the first beat of the bar. Be sure to keep the phrasing consistent with the original version of the music, or people will get confused trying to follow along with it, or even find that the power of the lyrics is lost. Stardust—'Music Sounds Better With You'—is an example where the vocal starts bang on the first beat of the section. But some music, like the example in the previous arrangement, uses a 'lead-in' on the vocal, where the first line actually starts just before the beginning of an 8-bar section. For an example of this listen to Rozalla—'Everybody's Free'—and hear how the main vocal hook places the word 'everybody's' *before* the start of the section, on beat three. Other songs occasionally begin their vocal slightly *after* the first beat, such as the opening verse in 'Happy' by Pharrell Williams, where the first lyric appears halfway through the second beat. If you were to import this vocal in with the first word at the start of a section, the phrasing would be completely ruined, with the emphasis on all the wrong syllables.

> Position them in the correct place. So many times I hear people start an acapella on the 1.1.1 but should come in on the 1.1.3 or before the bar starts etc.
>
> (Dave Winnel, DJ/Producer)

This is very tricky for DJs who are new to thinking about music on such a granular level. But if you get lost, listen to the original version of the song, and see if you can find the beat as you listen, identifying whether the vocal begins just before or just after the start of an 8-bar section, and where in the bar it goes. If you're still having trouble figuring it out, import the full original track into your mashup session, line it up so that it follows the 8-bar sections, *then* bring the acapella up next to it until it's playing over the top of itself.

BASIC MASHUP ARRANGEMENTS

Now that we have our elements imported it's time to really think about how the arrangement of our mashup will play out. With this in mind, there are a few typical song arrangements (also called structures) that always feel right on the dance floor. This is not so much a mashup principle as a general dance music principle. To achieve their intended effect, dance tracks need to feel like they are leading somewhere, and producers have learned not to confuse or disappoint their audiences with structures that don't respect this.

Mashups are supposed to feel like a single piece of music, and lead the listener on a journey, so the same applies. For this reason, the best way for you to start out is by following this arrangement of the main track in your mashup. Remember, whoever produced it has already worked very hard to ensure it works on a dance floor, so there's no need to change it up too much. The mashups we just discussed had little chance of failing because they simply stayed true to the arrangement of the main tracks they incorporated.

Arrangement takes a while to get used to. Watch your audience when you play your mashup out and see if they lose interest at any point. If they do, note it down. Later, you can go back to the studio and try something different.

> The first few times I play [a new edit], I watch the crowd to see if my arrangement is correct. Quite often I'll do a song and I'll see there's a point that doesn't work—it sounds right in the studio, but not on the dance floor. What might sound fine in the studio might be biting your ear off on the dance floor.
>
> (Phil Ross, DJ)

Given that most club tracks have at least some combination of an intro, breakdowns, beats (or drop) sections and an outro, we can look at some of the more common modern arrangements visually (see Figure 8.4).

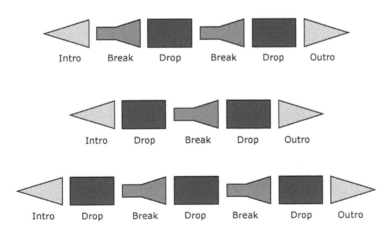

FIGURE 8.4
Typical Dance Music Arrangements

The first thing you probably notice is that none of these arrangements finish with a breakdown before going to the outro. To do this would build excitement and momentum in the breakdown, creating an expectation for an exciting section on the dance floor, but then disappoint the audience by abandoning the song instead and moving on to something else.

The exception to this is when a track drops out the beats during its final beats section and uses a very short 'break' section (around eight bars maximum) to give you one last hint of the main theme before moving into the outro. Rather than being a full breakdown, this is more of a strip-back 'sign-off section'. If it was any longer, the audience would feel that another beats section was coming up, and be disappointed. For an example where this is used to great effect, check out Sebastian Ingrosso & Tommy Trash—'Reload'. Notice how it leads you to the outro rather than winding you up for another big section.

Whilst this is a workable idea in fully produced tracks, I don't advise inserting it at the end of your own mashups because you will end up with three different songs over the space of 8–16 bars, which will work against the smoothness and subtlety that good mashups are about.

After the intro, whether the mashup begins with a breakdown or a drop will depend on two things: how much 'breakdown audio' you have available to you, and the intended length of the mashup.

If you plan out a mashup that is structured 'intro, breakdown, drop, breakdown, drop, outro', but you only have one single section of breakdown audio to use, things are going to get very boring for the audience, due to the direct repetition. If you want two breakdowns in your mashup and your breakdown track already has two breaks, or it is a song with two separate verse sections with differing lyrics, you're in luck. This gives the listener a sense of the journey of the original song. What is not as desirable (but still workable) is a chorus or repeated section of vocal over a slightly different bed of music underneath. Many club tracks already use the same 16 bars of vocal a few times during the song, but they are used over instrument sections that are a little different, and so they are still exciting to the listener each time.

In either case, both scenarios allow you to put two breakdowns in your mashup.

When you will run into trouble is where you attempt to use exactly the same piece of audio for both breakdowns in the mashup, because there is only one breakdown in the song you are taking it from. In this case, it is definitely preferable to start your mashup with a drop, and only have a single breakdown in your whole mashup.

The length of your mashup may also be affected by how safe you feel playing the track. An audience may be willing to listen to a monster-size 8-minute mashup of two songs they know all the words to, but a very unfamiliar mashup may need to be under three minutes long! Additionally, if you're going to make your mashup short, you should definitely go for the arrangement with two drop sections and a single break in between. As mentioned, avoid the opposite, i.e. two breakdowns and one drop section in the middle. Cutting down the dance parts in the mashup, and depriving them of a big section at the end of the journey will result in people feeling disappointed and walking off your dance floor. Another factor is to consider how your other mashups are arranged. If you make a new batch which all begin with a breakdown, the skeleton of every mashup will be the same. Just like regular dance tracks, some start with breakdowns and some start with drops. You should try and keep some variety, giving you more flexible moments and more varied DJ mixes during your set.

If you want to get really crazy, you can use two different main tracks as your two 'drop' sections. This has to be done carefully and tastefully, or the mashup will sound like it's trying to do too much, and start confusing the audience. Keep it tasteful, don't do it too often, and only do it when you believe it will really serve the audience. The two main tracks should be cleverly picked out, and arranged in an order that builds the energy rather than stepping it down. Above all, the moment of surprise when the music drops differently has to *feel good*.

Watch Video 2.2—'Mashup arrangement' on the website ('makegreatmusicmashups.com') to see some of the above arrangement principles being utilised in Live.

WHEN SHOULD I BRING THIS BIT IN?

One of the biggest hurdles you'll run into when you begin making mashups is being unsure what to do once you have everything sitting in the session. How do you figure out how to make one of the arrangements we just looked at? How do you decide what's going to work, or when it's a good time to bring in a vocal? One way to get started is by finding a mashup from a great DJ that uses the same types of elements. Listen through and see if you can figure out the reasons behind their choices—why did they start the acapella when they did? What tricks or techniques did they use to keep the mashup interesting? Did the mashup give you any hints that the music was about to change? Did it fulfil them?

Another way to learn is to just place an acapella over a track and see what it makes you feel. Does it build with the track? Does it feel strange when the beats drop out but the vocal keeps going, or does it feel good? Does that instrument coming in halfway through the drop clash with the vocal? Maybe I should just use the vocal in the breakdown. The main track already has some vocal samples in it—should I try and edit them out, edit out the acapella under them, put the acapella in a different section, or just chop out this whole part of the instrumental? Mashups are a bit like a puzzle, there are a variety of ways to put one together, but in order to serve one moment you may have to miss out on another. You have many options available to you, so it's important to think carefully about what the most important moment of your mashup will be, and arrange around that.

If you're looking at your mashup to-do list, you'll be able to see the thoughts you wrote down while you listened to the music. If you wrote a specific idea down, it's a good bet that it was a winner. Ignoring the rest of the arrangement, see if you can get the pieces to work together to create that specific moment. It might only concern the way that it kicks into the main track after the final breakdown, or the part where the vocal or breakdown track is revealed. Try creating it, and once you've nailed it, then have a think about how the mashup needs to be arranged in order to get to that point. In fact, see if syncing up that one moment has already laid your elements out in a workable arrangement. You may find that it's perfect the way it is!

When dealing with a long acapella that contains verses and a chorus, I find I like to bring in the verse over the beats to keep them interesting, then bring the chorus in at the breakdown, timing it so that the most climactic section of vocal occurs over the most hyped breakdown section. In this case, the big moment of the mashup is usually the transition as it goes from the big chorus into the biggest dancing section. In other mashups with short acapellas, I usually find I can bring it in during the breakdown, and carry it through to the big drop section, perhaps with some looping or chopping to keep it more interesting or to better match the energy of the drop. Experiment and see what you can find.

MAKING SURE YOU HAVE ALL THE AUDIO YOU NEED

Let's briefly look at an obstacle that may come up as you plan out your arrangement. At some point you will come up with a fantastic idea for a moment that requires a piece of audio you don't have. Examples:

- You need the most important vocal from the breakdown track to appear on its own, but there's heavy layers of music underneath and you really need to just cut to the vocal on its own in order for your idea to work.
- You have a pop track that has a nice empty verse section, but every time the chorus comes in a bunch of drums and big instruments fill up all the space . . . but you do remember you once heard a remix of the same song where the producer placed the chorus in a breakdown instead.
- There's a great synth or guitar riff used in a classic song, but in the recording, there's a drum fill or vocal in front of it that you wish you could get rid of.

In the first two examples, try looking for alternate versions. This is particularly useful regarding pop music; the genre that is mostly likely to be a) remixed by dance artists, and b) available as an acapella (either through official release or by extracting from a full version). I have been able to make many of my mashups better in those crucial moments by checking to see whether an acapella was available. Having the solo vocal to play with gives you hundreds more options and gives you the greatest chance of achieving the moment you want. It also means that the vocal can appear all throughout your mashup, and not just in the breakdown sections, which is great because it helps 'glue' the whole mashup together into a single piece of music.

You can perform edits that cut straight from a full song to its acapella version, so that from the listener's point of view the instruments effectively 'drop out' while having the vocal continue smoothly. If you do this, pay careful attention to the tonal difference. Usually an acapella will be lower in volume and less 'enhanced' sounding compared to how the vocal sounds within the fully mastered song. Use your ears and do any volume, EQ or reverb alterations necessary to make the acapella feel the same as it does within the full song. If the vocals match tonally, you can then perform an edit that cuts from the vocal breakdown straight to the acapella on its own, and to the audience it will sound smooth. This is a good trick when you want to place a lot of focus on the vocal for a couple of beats.

If an acapella wasn't released for the song you're working with, you may have a way of getting it anyway, which we'll get to in just a moment.

Regarding the second example, there have been many times where I couldn't quite get what I needed from a single version, and had to use multiple mixes of the same track and splice them to bits. I've particularly found this useful because a lot of dance remixes place the whole acapella (from verse to chorus) in the breakdown. This means there are no drums and no big bass-lines underneath! As long as the remixer has chosen to follow the same key and chord progressions as the original version, this can work brilliantly.

Be aware that different mixes of the same song often sound different, and that when using more than one, you should compensate with EQ. When remixing, dance music producers often add a lot more top end EQ and reverb to a vocal to make it sound more distant than it does in the original mix, so you may have to compensate by pulling back the top end EQ on the dance mix audio, or adding reverb and top end EQ to the original mix. Use your ears to match them however you can.

Lastly, when using two mixes of a single track within a mashup, be careful about instruments dropping out and being replaced by completely different sounding instruments as you switch over. If the difference is too great, it will sound strange to the audience, and you'd be better off finding an alternative solution by changing your arrangement.

When trying to find a single instrument riff, such a as a famous guitar riff or synth melody, do a web search for the 'stems' to the song, which can sometimes be available for really big tracks. (Stems are the individual elements of the track split up into separate wav files, prior to being mixed and mastered into the final track. This means you can find the element you want on its own.) It's a long shot, but stems (or remix parts) are available for songs that have been opened up for remix competitions, which can occasionally happen on big dance labels.

If you have producing or musical experience, you may also be able to recreate the riff yourself! If you are able to do this convincingly, it is an invaluable tool that will make you incredibly flexible, enabling you to create mashups that could not be made by anyone else.

'DIY' ACAPELLAS

As discussed in the previous section, sometimes you need a vocal on its own to create the moments you want. But, so many potentially amazing ideas require tracks that aren't released with an acapella. Though we try to make the best of it by using the full songs as breakdown tracks, sometimes we just wish we could get our hands on the vocal alone. The good news is that with a little luck and work, sometimes we can create our own acapella.

'Do-it-yourself acapellas' are the mashup artist's saviour.

Sometimes a track gets released with both a vocal version and an instrumental in the package. Generally, what this means is that you will have access to two files; one regular version with every element in the song playing, and one that's exactly the same only with the vocal muted. This gives us the opportunity to try and create a do-it-yourself or 'DIY' acapella.

First, let's discuss the concept of phase cancellation.

As we talked about in the Ableton Live chapter, at any given time the amplitude of an audio signal (which later translates to the position of the speaker cone) is somewhere between completely positive or completely negative. If there is nothing but silence in the signal, the waveform will rest in the middle, neither positively nor negatively charged (and the speaker cone will sit still in its natural position, directly in the middle).

When looking at audio as a waveform, the louder an audio signal is, the more wildly it swings positively (above the middle line) and negatively (below the middle line). It is this constantly swinging signal that Live displays visually to us as a waveform in the audio clip.

If you play two pieces of identical audio at the same time as one another, the peaks will be in sync. When these two identical positive signals combine together at the master channel, it results in a signal *twice as positive*, and the same for negative. Because the peaks are matching perfectly, the resulting audio signal as we hear it sounds just like one of the signals played with twice the power (+6dB in your DAW). This makes sense, because adding two identical signals together should be the same as just doubling the power of one of them (Figure 8.5).

But, if we 'invert' one of the Audio Tracks, we flip the wave around its zero line, converting all of its positive peaks into negative peaks, and vice versa. When one audio signal is combined with another that has an identical *but inverted* signal, all of the positive wave peaks of one track will be combining with the

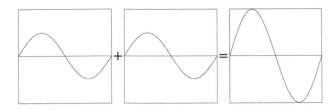

FIGURE 8.5
Two Identical Waveforms Being Added Together

exact negative peaks of the other, resulting in a completely *phase-cancelled*, silent piece of audio.

We can use this to our advantage when we know two audio files to be identical, except for one small difference: one doesn't have a vocal in it. Using the above theory, if we line up the wave peaks *exactly*, and invert one of the files, all of the audio information common to both signals will cancel out, leaving only a solo vocal, with almost or absolute silence behind it (Figure 8.6).

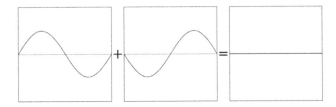

FIGURE 8.6
Two Identical Waveforms, One Inverted, Being Added Together

We can create this phenomenon ourselves by creating a new Live session especially for the task. Import the vocal and instrumental versions of one song, placing them on neighbouring Audio Tracks. Don't perform any Warping on the audio; make sure to turn Warping *off* on both clips. Concentrate on creating the best isolated acapella you can; worry about how it will fit into your mashup later. First, find out the tempo of the song (you can use the Warping function to help you, but make sure you turn it back off afterwards) and set your *session to this tempo* to assist you in making edits later, if necessary. Zoom right in and make sure the peaks between the two tracks line up *exactly*—don't assume the original and instrumental mixes have the same amount of silence at the start (Figure 8.7). To drag the audio around freely on the grid, go to Options and turn off 'Snap to Grid'.

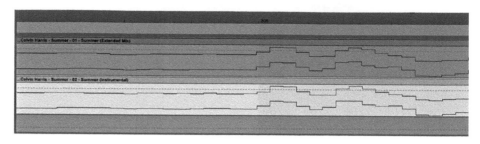

FIGURE 8.7

Lining up the Original and Instrumental Versions

Screenshot Appears Courtesy of Ableton

Use the beginning of the tracks to sync them if you can. However, if the tracks don't begin with drums, it will be easier to use a later section where there are easily identifiable transients such as kicks and snares. Bring the two Audio Tracks down to –6dB so that the audio won't clip at the master. Then, to invert the audio signal on one track, place a Utility on the Device Chain for the instrumental version of the track, and enable both 'Phz-L' and 'Phz-R'. Have a listen through. If the versions are truly identical except for the vocal, there should be little or no music or drums behind the vocal. Now that you're able to hear what the phase cancellation is doing, you can subtly move the instrumental track around and hear where the best spot for cancellation is. Sometimes one element will sound more cancelled at a certain sync point, but another starts to creep through. This is ok, just find the best balance you can. I suggest using headphones for this part, as tiny sounds and high-frequency bleed-through can escape your attention on speakers.

Be aware that even the final master of a song can sometimes drift, and you will need to listen through and make sure the tracks line up all the way. If they don't, create small splices in one of the tracks where the peaks start to go out of sync and shift it manually back into sync. Try to time your edit points between lyrics, because any sections without vocal can just be muted later. Also bear in mind that instrumental versions are often 8–16 bars shorter than the original song, edited down so as not to get boring (Figure 8.8). If so, you will need to listen through and figure out which sections have been removed from the instrumental so that you can splice and move the clips accordingly, and make sure the right sections sync up. This is a good reason to have your session and grid set to the song's tempo.

Do the best you can to make sure the instrumental elements phase-cancel as well as possible, but don't stress out if your **DIY** acapella is not pretty. It doesn't have to be perfect, just usable. We're really at the mercy of what process the producer went through when exporting their song. Anything created via random fluctuations, such as the oscillation of synths, white-noise generator effects or the random scattering of reverbs will differ between versions unless they were

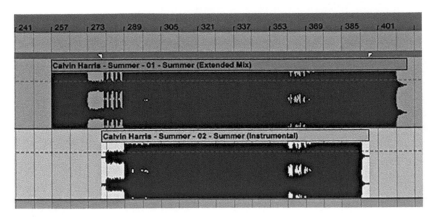

FIGURE 8.8
Instrumental Versions Can Sometimes be Shorter

Screenshot Appears Courtesy of Ableton

rendered at some point during production. Such sounds will not sync up peak-for-peak between versions, and you may hear some drifting noise or effects sounds in the background of your finished DIY acapella. There's nothing we can do about that. Although quite ugly when heard on their own, these little musical ghosts are usually not a problem in mashups. The producer has timed white noise effects to occur at the same points we want them to anyway, and extra reverb and synth noise will at least be in key with the rest of your mashup. Even a tiny bit of drum information leaking through can be ok, because you'll be making sure the vocal is in rhythmical sync anyway.

After you have decided your acapella is as isolated as you're going to get it, get to work on cleaning it up. First, you need to use muting and fading to completely silence the parts of the acapella that aren't useful to you. I often render what I have at this point so that I can work on a single wav clip, but I suggest you use Track Volume automation on the master channel, in case you find little sync issues you need to go back and change (Figure 8.9). Use the automation to completely silence long sections between vocal sections, and to perform little fades in or out of vocals to help hide those little musical ghosts as lyrics taper off.

Next, use EQ to clean up the vocal and hide anything around it. It is important only to use EQ on the master channel: on the combined signal. If you use EQ on one of your individual tracks, it will actually change the phase relationships and push the peaks around, causing the two tracks to come back out of sync. Select a section of vocal where they are singing at their lowest pitches, usually a verse, and loop it (CTRL+L, CMD+L). Create an EQ plugin on the master channel, setting up a high-pass filter with a subtle roll-off or Q. While playing, start from the lowest frequency and slowly pull it up until you notice the bottom end start to disappear in the vocal. For males, this can occur somewhere

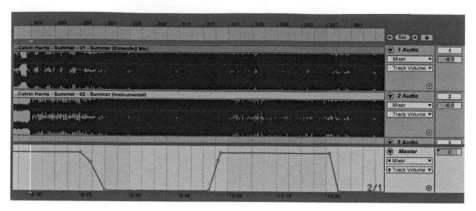

FIGURE 8.9
Using Volume Automation on the Master Track to Clean Up

Screenshot Appears Courtesy of Ableton

between 70 and 100Hz, depending on the range of the vocal. It is usually higher for female vocals. Once you hear the bottom end start to disappear, back it off by about 25 percent; i.e. if you hear the vocal being affected at 100Hz, back your filter off to 75Hz to be safe. This should take care of any thuddy remainder of the kick drum and bass, as well as unnoticeable low-end audio problems as a result of the phase-cancel process. Be conservative, you can always add a more severe filter when you're working on your mashup.

If there is a little bit of high-hat noise spilling through, you can try the same thing using a low-pass filter up the top. Bring it down from the highest frequencies to see whether you can eliminate some of the hats. But be very careful, it's much more important to keep the sibilance intact for lyrical clarity than it is to silence drum spill. Acapellas are useless if you can't actually hear the lyrics. If in doubt, don't alter the top-end at all. If you think any additional EQ band changes around the middle can help clean up the vocal, go for it. There may be a snare or high-hat poking through that you can use a very surgical notch EQ to pull out. Just be sparing with them, because they're likely to negatively affect the vocal.

Lastly, add an additional Device or plug-in that has mid-side functionality. Waves 'Center' is a good one, but many EQs will also provide this control. EQ Eight does, so feel free to use that. In this instance of the Device, we'll switch the Mode from stereo to M/S (mid-side). When set to the default stereo mode, any EQ change applies an identical change to both the left and right channels of audio. Setting it to L/R mode allows you to make independent changes between the left and right channel, for instance pulling out a frequency on the left side, but leaving the right as is. Similarly, mid-side mode allows you to independently control two channels of audio—everything that is common to the centre of the stereo field, and then everything that sits at the sides of the stereo field (audio not common to both sides). The plugin performs a conversion to work on the audio in this way, and re-converts it to stereo at the end, so that the signal sounds normal to us. The reason we want to process the audio in this way is that the vocal almost always sits directly in the centre of a stereo mix. Remember those elements we always have problems with during phase-cancellation? White noise, reverbs and synths—they are almost always spread out across the stereo field. Using mid-side processing, we can make changes to (and hopefully eliminate) these elements while leaving the vocal relatively untouched.

First, pull out absolutely everything on the side to see whether this improves the clarity on the vocal. If, like EQ Eight, your EQ doesn't have a volume control for each channel, do this by selecting 'S' or selecting 'side' mode, then setting up a high-pass filter and literally setting the frequency to the highest setting (Figure 8.10). If you play your audio, you should feel all the stereo information get sucked out of the music, leaving you with audio that sounds like its coming from directly in front of you.

This may be all you need to produce a noticeable change in the acapella, dropping out a lot of the extraneous musical ghosts. If you find that the vocal

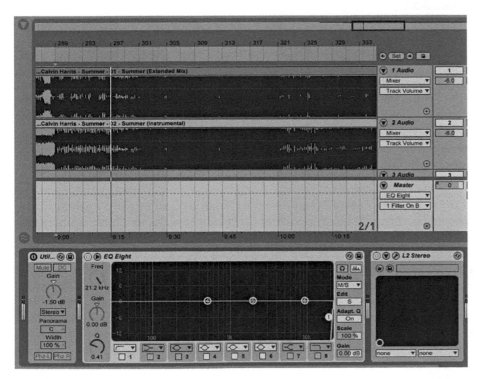

FIGURE 8.10
Using EQ Eight to Eliminate 'Side' Information

Screenshot Appears Courtesy of Ableton

itself suffers a lot, it may be because the vocal is double-tracked (recorded twice, each version panned slightly to the side). If this is the case, you will have to settle for pulling out perhaps 50 percent of the side signal. If your EQ doesn't have a volume controller, change your high-pass to the low-shelf setting and, starting at 0 gain, pull it down until you find a good balance between the instruments disappearing and the vocal starting to lose its integrity (Figure 8.11).

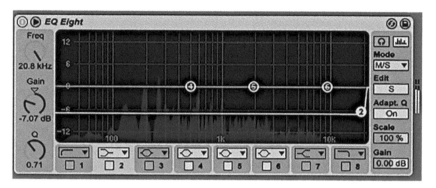

FIGURE 8.11
Using EQ Eight to Partially Attenuate the Sides
Screenshot Appears Courtesy of Ableton

Listen through the whole acapella at this stage to see whether any sections have slipped through the net. Switch between using headphones and speakers, to make sure you hear detail in the headphones and low-frequency problems on the speakers. In the Device Chain, make sure there is a limiter at the end. Just use the same limiter and settings you would use for your mashup template. Make up the 6dB loss we began with, either with a utility plugin before a Limiter Device, or within the limiter Device itself.

To finish up, go to options and re-enable 'Snap to Grid'. Create a selection on the timeline (on any track) to define the area you want to export, and render your new DIY acapella to a wav file of its own, where it can be used in mashup sessions later on (see the section on exporting at the end of Chapter 9). Ensure that you begin on a bar that makes sense, to make things easier for yourself when you import it later. If there is a lead-in to the first lyric, you can even provide a whole eight bars of silence beforehand, just to be safe. When you export it, make sure you specify in the file name that it's a DIY, to warn yourself to listen out for those effects, floaty background noise or the subtle remains of drum loops. You should know the tempo of the vocal now, so include this in the file name too, to save you having to figure it out every time you use the acapella from here on.

This method can get you the acapellas for some seriously good songs, and really open up your options for creating great mashups. It's unfortunate, but less and less music is released with acapellas and even instrumentals these days. It used to be standard for dance tracks to be released as an Extended Mix, Original (or radio-style) Mix, and an Instrumental Mix. With this in mind it becomes more and more necessary to try and create acapellas without the aid of an instrumental.

Though they won't usually sound very good on their own, some DJs make *very usable* DIY acapellas even when there is no instrumental track to phase-cancel with.

There are two ways to go about this, and both use the techniques explained above.

First, if you are trying to grab an acapella from an electronic-based track with plenty of repetition, you may be able to use the above method exactly. Instead of placing an instrumental version on the phase-flipped Audio Track, you simply place a section of the original that has no vocal over it. You use the same syncing method to line it up with a section of the music that *does* have vocal in it, and see what cancels. If the two sections are instrumentally similar enough, you might be able to get a passable cancellation. At the very least, the drums will usually cancel nicely, perhaps leaving a layer of high-hat, ride cymbal or a small layering synth, which is not a big deal. You'll have to do a lot of chopping and changing to find out which parts phase-cancel the best, which is why it's important to have the grid correct in relation to the song.

Sadly, this technique requires the music underneath to be very simplistic, repetitive and to contain lots of instrumental section. Many tracks simply aren't set up that way anymore; instead they using short, ever-changing sections to keep audiences interested. The biggest obstacles are changes in the musical elements as the song progresses, such as fluctuating chord movements and constantly varying melody lines. These will not cancel when sampled from different sections, instead summing to create multiple musical lines smashed together and becoming twice as noticeable as they were before. But it's worth a try, if the acapella is *seriously* desirable. It's a little more likely to work in either older tracks or drawn-out styles such as house music, which will often only use a short 8-bar phrase of vocal. The changes in this style of music are also so patiently spread out across the track that the addition of a vocal is likely to be the only change in that particular 8-bar phrase. This means it will cancel beautifully!

The second, absolute last-resort method is to literally apply processing to the track without using any phase-cancellation. This pretty much never gives you an acapella of the quality needed to do remixing with, but if it's used in the right context, it can just *barely* scrape across the line in terms of mashup usability.

For this method, you just import your song and begin by using the EQ and mid-side processes you learned just now. You'll depend very heavily on your mid-side processing for this. Many DIY acapella creators also use volume automation, or even complex plug-ins for automatically muting audio below a certain volume (called noise gates) to pull the volume down in the small gaps between vocals. This is a tricky one though. I believe that it often is more trouble than it's worth, as the listener will hear strange foreign instruments launch into awareness every time the vocalist sings something, before disappearing just as suddenly. This draws even more attention and causes even more distraction! In cases such as this, it's usually best to leave the extraneous instruments running at a constant level, and to only use the acapella in mashups with tracks that can effectively obscure those instruments. One example that comes to mind is the acapella for 'Hey Boy Hey Girl' by The Chemical Brothers. There was a DIY acapella floating around online that I was always tempted to use, but the filtering and gating that had been necessary to hide the drum loop and instruments was so devastating to the vocal that it sounded like someone murmuring through a telephone line, as well as lurching the backing track in and out each time the vocal played. Instead, it was better to wait for a mashup idea where the instrument and drum loop would match the vibe in the mashup, so that I could use a less demolished version and allow the audience to hear the lyrics in perfect clarity. If in doubt, resist the urge to filter or mute too harshly at this stage, because you can always do more of it later, on a mashup-by-mashup basis.

Regardless of which method was required to produce your acapella, after you have exported it, you can still try one more technique to get those last remaining ghosts out of the audio. You must always make this the last thing you do though, otherwise it can prevent the other techniques from working. For many years now, the more specialist DAWs and audio manipulation programs have provided noise reduction processors in order to reduce hiss on old recordings. On forensic-investigation-themed television shows, this is how they remove the sound of passing cars from CCTV recordings to hear the murder victim's final words! They work by feeding the noise-reduction processor a sample of the noise on its own, so that it can build a profile of the frequencies that make it up. It then reduces those frequencies across the whole recording in an attempt to remove the noise.

For a free plugin that does the job, look no further than the Cockos plugin, ReaFir. Designed for their DAW 'Reaper', it is available free from www.reaper.fm/reaplugs/ and works as a noise-reduction processor when set to 'subtract' mode. For paid options, 'denoise' in iZotope's 'RX' or 'noise reduction' in Adobe's Audition are great options. Select the largest section of the music that you can find that has no vocals, but all of the left-over noise and ghosts that you don't want. Then open up the noise-reduction process and tell it to 'learn' the noise profile you have selected.

Then select a short section where the vocal plays and experiment with the settings, previewing each time. Too much noise reduction will reduce the frequencies it interprets as 'noise' so much that it will affect your vocal, usually resulting in dulling and a strange morphing feeling in the vocal. Back it off to an acceptable level. Also, play around with frequency settings, if there are any. Reducing the noise reduction in frequency sections where the process is doing too much damage, such as the top end, can rescue the lyrics. As you'll hear, the processor faces a pretty tough job trying to reduce the noise without killing the vocal, because the human voice covers so much of the frequency spectrum. But a little bit of it can help. You'll generally get much better results if you were able to phase-cancel away much of the instrumental first. Once you've found your desired balance, extend your selection to the whole file and process it. Listen through closely, and if you're happy—export it.

Lastly, I want to stress that you shouldn't feel disheartened about the lack of acapellas available today. Though they have been a great mashup tool in the past, you can still take advantage of a great vocal simply by using the song as a breakdown track, and by utilising smart arrangement to put the vocal in the places you want.

Head to the website ('makegreatmusicmashups.com') and watch video 2.3—'Making do-it-yourself (DIY) acapellas' to see the above techniques in action.

GENERAL TIPS ON ARRANGEMENT

Before we go into specific techniques, here are some general tips for the arrangement of your mashups. These kinds of ideas will eventually come automatically to you, so don't stress about memorising each one!

■ Remember to use space and avoid two new elements coming in at once. Let the audience get used to one before you bring in another. For example, if you've just transitioned to a breakdown, try to let it play for 4–8 bars before bringing in a vocal. If you've just transitioned into a drop, let the audience get used to it before you bring in a vocal or something else.

■ Related to the above idea, try specifically not to bring in a vocal right at the beginning of a break or drop. A vocal is usually an important element and requires the audience's full attention. So again, they need time to feel the new breakdown or kick-in before they're ready to focus their ears on lyrics.

■ More often than not, you need to use a reverse reverb to bring in a vocal, to help the ear compensate for the fact that the vocal does not naturally occur over the song beneath it. More on this technique later.

■ If you decide to have two breakdowns in your mashup that both use the same 'break track', make the second break interesting, don't just repeat the first. If the two breaks are too similar, do something special in the second break, like introducing a vocal towards the end, or including a drum loop over the break. You can make it eight bars longer, or eight bars shorter. Play with the expectations of the audience. Try to have it transition back to the main track in a slightly different way. Better yet, if you're using an acapella, avoid using it over the first breakdown's build-up, holding it off until the second break's build to really bring it home.

- For mashups that use a vocal, try and use the most catchy or well-known hook just before the main track kicks back in. Even if this means dropping out some or all of the elements behind it to bring focus to it.

- You may choose not to use some sections of a vocal, either to fit it in to a short section or prevent the mashup from becoming boring. If you do this, be careful to listen to the lyrics and make sure they still make sense. It's easy to get caught up in how your mashup is working and forget about the story behind the vocal. In some songs, the chorus is the most important or well-known section. In others, it's the opening verse.

- Always be careful how much you crowd into a mashup at once. If you try to play multiple elements together that are too full, you can create trouble. Try to avoid it sounding messy, or the audience will get confused about what to listen to. If in doubt, less is more.

- If you want to put a lot of focus on a short section of vocal (from one beat up to one bar long) try muting or high-pass filtering away the main track underneath it momentarily. This works well when using a short vocal hook during a main-track section, and you want to draw attention to the vocal without the beats and bass-line trampling all over it.

- If you find you need more energy in the final section of your mashup, gradually pull the master volume down around 1dB during the build-up, then slam it back up to full—right when the main track comes back in. It is only a subtle change when listening in your studio but it makes a difference when heard live.

- Listen to the mashups in sets from your favourite DJs. Write down everything that happens as it plays through your speakers. This will teach you *so much* about good mashup arrangement.

- Lastly, experiment!

EDITING INTROS

> Mashups and re-edits help to keep interest in a culture that is growing more and more used to a quick turnover of music.
>
> (Phil Ross, Commercial DJ)

Once you feel you have come up with the best arrangement plan for your mashup, start putting everything in place.

You can either start from the most important moment of your mashup, or you can start from the beginning of the first element. If you're working on the first section of your mashup, it will generally be your main track intro, unless you're starting with something unusual such as a set intro or something you don't need to mix into. While you have the audio in front of you, this is a great opportunity to edit your intro to a more desirable length. Your reason for doing this is that songs from different styles and eras conform to different standards. For instance, more underground styles such as tech house or extended mixes from the 90s are more likely to utilise an intro of 32 bars or more, equating to over a minute of music. On the other hand, more commercial and fast-paced styles such as subgenres of electro can use intros as short as 8 or 16 bars, particularly styles where the producer anticipates that the DJ will employ a fast turnover of music. As dance music audiences get more used to fast paced sets, DJs find they have to mix between songs quickly, so intros are getting shorter all the time.

Given that sets are more rapid than ever before, many DJs like to be able to pick and mix a track very quickly during a set, and shorten the intros on all their music to 16 bars (around 30 seconds at 128 BPM). Other DJs like to extend theirs out to 60 seconds or more so that they can do more creative mixing and let the audience enjoy the transition. As long as you keep your intro length to a multiple of eight bars to make sure you can still bring it in at the right moment, you can shorten or extend it to however long you like, depending on your style as a DJ. If you need to extend it out, try and time your 'pickup' moments to points that make sense. For instance, if you are extending a 16-bar intro to be a 32-bar intro, don't give the audience eight bars of low energy, followed by 24 bars of higher energy; give them 16 bars of each. Listen closely to the elements within the production. If the track uses building effects such as snare rolls and white noise risers to build to the end of their intro, it is going to sound strange when you repeat that 8-bar section. The audience willfeel like the track is about to drop in, then suddenly go back to a smaller energy and build up all over again. In a case like this, it actually is better to repeat the first eight bars three times, and only then allow the intro to continue to the building section, while using your own build effects and elements to try and aid the build through the first 24 bars. You're going to have to use your ears constantly to make sure you don't create any jarring surprises for the audience.

On the other hand, if you are shortening a 32-bar intro to be 16 bars, you need to be careful which 8-bar sections you choose to remove. If you pick the wrong one, you'll end up with massive jumps in energy, which will sound disjointed and confusing to an audience. Sometimes, the rise in intro energy is so constant and gradual that there is no 8-bar section you can remove without causing problems. In this case you can either remove your 8-bar sections from the very start of the track, leaving the later section intact, or you can perform a fade.

Using a fade to smoothly skip an 8 or 16 bar section can be very useful, but you have to make sure you get it right. First, you must only perform it on audio that is *not Warped*. You do it by selecting 16 bars of audio, splitting it into two clips, and performing a gradual fade so that over the course of eight bars, the audio fades from the first clip to the second, effectively covering 16 bars of the track's original energy build, but having it occur in the space of eight bars. You'll need to splice your intro in the middle of the 16 bars, and move one of the clips to a second Audio Track. Shift your clips so that the sections that must overlap are doing so. See Figure 8.12.

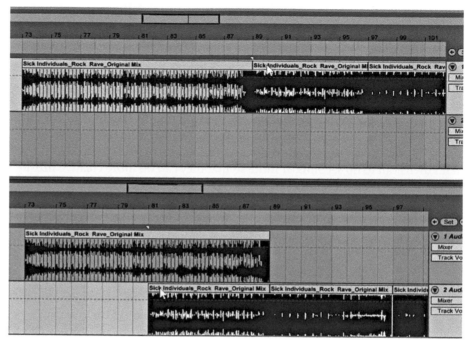

FIGURE 8.12
Overlapping 8-Bar Sections to Shorten an Intro

Screenshot Appears Courtesy of Ableton

Zoom right in to make sure all of your beats match up *exactly*. This is hugely important, because if they don't match up sample for sample, the result will be a washy, flanging effect as the two almost identical pieces of audio play ever so slightly out of sync (Figure 8.13). This is why you cannot use Warped audio, or the peaks will drift around and you will hear *flanging* and *flamming*. Hold Alt (Win), CMD (Mac) to bypass Snapping to Grid, allowing you to freely drag the clip on the arrangement page.

Activate the Mixer > Track Volume automation lanes on both tracks. During this intro section, if only one clip is playing at a time, you want it to play at 100 percent of its natural volume, which is 0.00dB on the volume fader/automation point. When there are two clips of the same song playing at once, you need to bring one or both of them down in volume, otherwise the elements common to both clips (usually your drums) will playback twice as loud, or +6dB compared to one playing by itself. The limiter will not like this, and while trying to squash this excessive audio down to an acceptable level, the music will crack up and sound bad.

Try to have one of the Audio Tracks playing at −3dB automated level, or settle for a compromise and have both Audio Tracks playing at around -2dB. If one of the clips is a section that doesn't have as many drums in it (usually the very

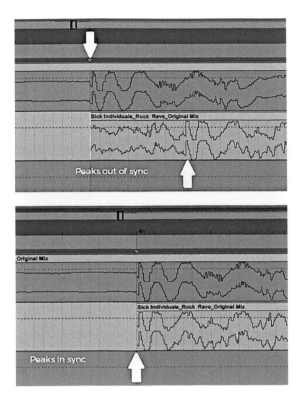

FIGURE 8.13
Lining Up the Peaks
Screenshot Appears Courtesy of Ableton

end of the intro) you can allow more of it in. As always, use your ears. You want to listen for a good balance between the two audio streams, but also listen for any uncomfortable drops in volume. The aim is to be so smooth with your fade that the audience doesn't know eight bars have been bypassed. If you hear any squashing and distortion, alter your automation to make it comfortable. Figure 8.14 shows an example where an intro of four 8-bar sections is reduced down to two 8-bar sections by fading from the second 8 to the fourth 8, and completely skipping the third 8.

Have a close look at Figure 8.14. The final bar of the second 8 is also completely muted to make sure there's no drums left underneath the vocal that appears in the last bar of the intro. When editing an intro like this, be aware of any special moments in the intro of the song you're using. If you're performing a fade between two clips, but one of them intentionally drops out all the drums and effects one bar before the kick-in, you don't want any lingering audio coming from the other clip, so splice it at the same time the other clip drops its drums, and delete/mute it. If the producer has intentionally dropped out the drums before everything kicks in, you don't want to destroy the effect.

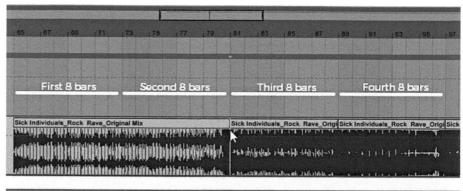

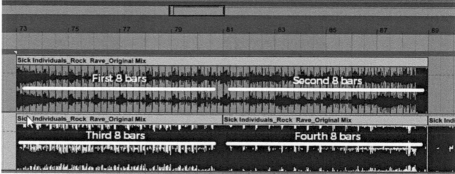

FIGURE 8.14

4-Bar Sections Being Shortened to 2 with Splicing and Crossfading

Screenshot Appears Courtesy of Ableton

Sometimes you will need to silence a clip in this way just for the space of one single beat. Big room tracks often place a giant snare sample on the last beat before a kick-in, and cut out everything behind it. House tracks often provide a half-bar or full-bar of drum-fill with nothing behind it. Similar to the previous example, you need to preserve that effect and make sure there's no smaller drums or residual effects hanging around in the background coming from a second fading clip.

If you wish, you can use all of the above techniques for editing and altering the length of outros as well. If you need to extend a track with a very short outro (eight bars) it is often quite difficult. Sometimes, you can just repeat a single bar rather than a whole 8-bar section, and duplicate it out to create your own 8-bar section. Usually this works best if there are no effects sounds in the outro, but just simple drums. Finally, if you need to extend an incredibly short outro, see if there's eight bars or a single 1-bar loop in the intro which you can copy to the end. As long as it goes smoothly from one to the other, it's fine. Much of the time, the intro and outro of a song are made up of almost the same simplified elements. Just use your ears and make sure it doesn't sound like anything new is coming in during your borrowed section, or this will confuse you and the audience when you're mixing from this track into another during your set.

Watch video 2.4 'Editing intros' to see how the above technique is done.

CHAPTER 9

Making Mashups—Part 2

Techniques

So far we've learned that it's crucial to really think about what we want our mashup to achieve, what moments we want to create, and to think about the arrangement we can use to do it. The above points are great guidelines, and if you have picked good ideas using elements that are reasonably compatible, you will already be on your way to making great mashups. In fact, if you've picked an absolutely brilliant pair, all that may be required is a decent arrangement and a few fades! But most of the time, it's not that easy; we have to do a bit of work to make them blend properly. Now let's move on to specific techniques we can use to make our elements work together and keep things sounding smooth and professional.

> It isn't always easy to learn the following techniques without hearing them. It's a great idea to jump on the website ('makegreatmusicmashups.com') and follow along with videos 3.1–3.11 to see how and why they work.

TRANSITIONS

Now we come to one of the most important ideas behind great mashups.

Since we are trying to create a single 'track' that changes from one song to another multiple times, it is essential that we become good at making smooth transitions. Most of the following techniques in this chapter are designed to help you do this. There are a few things to consider for each transition you perform:

- length (how long should it take to transition from one song to another);
- point of realisation (when we want the listener's ear to realise that it has switched to another song);
- change in energy (whether the energy gets bigger, smaller or stays consistent).

The length of a transition depends on what kind of transition it is. The more sudden and surprising transitions may cut from one to another instantly or over the space a few seconds. These can be exciting, but if not done well, they can sound awkward and confusing. Longer transitions that aim to be smooth and subtle may take 16 bars or more. When executed correctly, they can create great momentum and smoothness, but if not, they can sound slow and boring. The most common transitions sit somewhere in the middle, over the space of an 8-bar section. Remember, when hearing club music, our ears are trained to think in 8-bar sections; we instinctively *know* that roughly every 12–18 seconds, an opportunity arises for the song to *do something*. Whether transitioning from a breakdown to a main track, or a main track to a breakdown, a length of eight bars feels natural and is usually the perfect balance between surprise/excitement and smoothness.

Somewhere between the beginning and end of this transition, we want the listener to realise that something new is happening. This generally occurs when the presence or loudness of the new element becomes greater than the first. This is determined by the speed and shape of the automation you draw into each element's volume and filtering **envelopes**, which we'll look at in detail shortly. In an 8-bar section, this crossover point may happen somewhere around the middle, but don't be afraid to have it occur earlier or later depending on whether you want your listener to feel the change coming with plenty of notice, or for the new element to be more of a surprise. If you intend to do 8-bars of musical transition as well as bring in a vocal near the end, you will want to spread your two changes across this 8. Time your musical crossover to occur a little earlier than the middle point so that the ear has time to adjust, then bring in the vocal.

The aim of a mashup is to give the audience multiple songs combined into a track that still feels like one single piece of music. Sometimes the energy needs to rise up a little (such as when it transitions from breakdown to main track) or settle down a bit (when going into a breakdown or slowing down in tempo). But it is still important that any change in energy is intentional and gradual. The ear notices unnatural jumps in energy and draws attention to our edits, so during our transitions it is important to keep the energy level reasonably stable. This is more crucial as the transition lengths become shorter. If you perform a transition of four bars or less, you'll have to work your volumes and filtering very hard to make sure there is no sudden jump up or down in energy. Relating to this, you need to make sure that the energy in the middle of the transition (particularly at the crossover point) doesn't feel too high or low compared to the sections either side of it. You want it to feel like one smooth line of momentum.

By the way, don't rely solely on your master limiter Device to control big peaks of energy during your transition. Overall energy and presence doesn't just depend on volume, but also the filtering and bass balance between two tracks, which the limiter can't control as effectively. Besides this, the squashed and over-compressed sound that occurs when overloading a limiter is perceived by the ear as loudness, so the energy will still feel unstable. Make sure everything is smooth *before* it reaches the limiter.

BRINGING IN AN ELEMENT

How you bring a new element into the mashup is dependent on your style, the type of music you're using and how you want to manage the expectations of your audience. In a way, a mashup is just one sequenced unveiling of ideas—some things are revealed slowly, some spectacularly, some understated and some unpredictably.

The simplest way to bring something in is just to start playing it—or by bringing it in 'cold'. Most elements within a dance track do this anyway; drum pickups, instruments starting or a vocal coming in. It is the most bold and honest way of bringing in a sound. However, because mashups are often trying to transition from one sound to another, not just add more in, we have to use gradual changes to hand the illusion of rhythm and key between songs. Because the dance floor loves 8-bar sections, this will usually occur over the space of 8 or 16 bars. Some sounds can be volume-faded in. If you don't want the listener to really know that the element is coming in you can choose to sneak it in very gradually. Just be aware, it has a similar feeling to the moment when a DJ crossfades from one song to another in the club. Keeping this in mind, a fade doesn't always sound intentional, and doesn't necessarily make the elements feel like they're part of the same song. A more classy way of bringing in an element is the use of a low-pass filter.

If you want the listener to hear that something new is coming up, but obscure its clarity to begin with, use low-pass filtering. The added advantage of filtering in is that as it gradually 'opens up'; as its target frequency ascends it contributes a feeling of building to the section. When the element is finally revealed in full, dropping other elements out behind it doesn't feel like so much of a let-down. The resonance or Q on the filter also plays a part—a high resonance creates a lot of focus on the cut-off frequency, directing attention to the movement. Low resonance allows it to happen more subtly, so that the listener might not realise anything is happening until halfway through the 8. Low-pass filtering an organic element in, like a rock song, also helps it to feel more 'DJ', more electronic; fitting it into the concept of the mashup more naturally. Often though, introducing an element via low-pass filtering requires some amount of volume automation in addition to making sure it feels even. Also, don't be afraid to use your EQ to pull all the bass back on the new sound until you've finished filtering it in, and have completed the transition—particularly to avoid bogging up the bottom end with resonating bass notes or drums. Do whatever you need to do to make it fit.

The other type of filtering, as we have learned, is high-pass filtering. You don't hear a high-pass filter used much as a method of bringing a sound in; usually a way of transitioning out of it instead. If you start an element with the high-pass filter all the way up; the high-hat, vocal sibilance and top end from the instruments and whooshes will suddenly be present in the music, but it will be seriously lacking in punch and foundation until the very end of the transition. When we bring an element in, we usually want it the other way round. It will also create a feeling of falling instead of rising, as the cut-off frequency moves downward. Still, if you want to hear an interesting example of a song that begins by high-passing in the first element, check out 'Flawless' by The Ones.

The method you use to bring an element in is crucial to how you manage the expectations of the listener—do you want to surprise them with boldness or use subtlety to create hints of what is to come? Is there too much musical information in the element to have it come in cold? Do you want to low-pass filter in a looping section of something and *only then* allow the section to play in full?

Another great way to experiment with how filters feel is to just loop a section of a mashup you're working on and play with it live using the mouse or a MIDI keyboard controller. This will teach you a lot about how filters feel, and how those feelings can be used to your advantage to shape transitions.

HIGH-PASS FILTERING DURING TRANSITIONS

The business of high-pass filtering to transition *out* sounds is so crucial that we should look at it in detail. Every mashup you do will require some (if not lots) of it.

The dance floor loves taking steps forward into higher and higher intensity; it does not like taking steps back. But how is this possible? If a clubber goes and spends four hours in a nightclub, they're going to hear between 700 and 1000 8-bar sections pass through their awareness. How can each one be bigger than the last? Part of being a dance music producer is basically 'tricking' the audience in a way where they don't feel they are being cheated! Producers avoid fading things down, avoid low-passing things out, and when they drop something out cold, they substitute big sounds for new, little sounds they can build from. Like Indiana Jones, they always slip something into the audience's awareness to replace whatever they have stolen away. This is pretty much why breakdowns work, although they in themselves have to immediately continue the trend of building too.

So that's the crucial concept—when we drop something out cold, we must replace it with something new to keep the audience excited. What about when we transition it out slowly? Using a high-pass filter to slowly remove it from the mix is the smoothest way to get rid of it. The energy itself stays at full power; the high frequencies remain playing, keeping the *intensity* going, and therefore not disappointing our little dancers. As the filter frequency moves up, the feeling of rising continues to excite the dance floor and lead them into what they interpret as higher energy. But down the bottom, the lower frequencies filter out bit by bit, diminishing the presence or proximity of the sound. Of course, by the time the 8 finishes, we had better be ready to substitute something new in.

Conversely, the other methods such as low-pass filtering or volume fading pretty much always result in a noticeable fall in energy and an awkward transition. It can even ruin a big kick-in moment that comes afterward because the audience aren't given any warning to expect it.

To 'high-pass out' an element, simply activate the Device at the beginning of the section where you want the element to filter out. Starting from the bottom, automate the frequency value to rise throughout the section until the sound has sufficiently disappeared. Remember, you don't always want to high-pass the element out completely; having the filter rise to around 80 percent of its range will still allow the intense high-end frequencies to reach the listener, a perfect idea for the end of a build-up section. While the intensity continues, all of the lower and middle frequencies are filtered out, and the element will give the listener's ear almost no foundation or key information, which leaves their perception open for influence from the next incoming element.

After you have set a basic automation path for the section, go back and listen to the smoothness of the filtering, and make any adjustments needed. You may find you need to smooth out some sections when particular elements such as kicks, basses or vocals disappear too suddenly. Also keep a close eye on the balance between the first element and the second; not only making sure that the first disappears nicely, but that the whole transition occurs smoothly. If you can't make a transition sound smooth using filtering, you will need to try something else, or reconsider the two sections or elements you are trying to blend.

Remember to keep the crossover point between elements in mind when you draw your high-pass filter automation. The amount of bass being filtered out of your sound makes a huge difference to the amount of 'presence' it has, and is just as important as volume when balancing your elements. A *subtle* amount of volume fading can be used when high-pass filtering an element out, but just be careful that the energy levels stay smooth. Mostly it should be used to smooth out unnatural lurches in volume as the resonant filter sweeps past frequencies containing prominent notes or drums.

During the sections where you aren't using a high-pass filter, it should be switched *off*. Use automation to have it switch on when you need it, and switch off when you are finished with it. The reason for this is that high-pass filters create a bump in volume around the cut-off frequency. When this frequency is set to its lowest value, you won't hear this bass bump with your ears, but it will cause problems with your limiter and also make a mess of things when played out in a club, causing the volume to jump around unexpectedly as the club's own limiter system catches big bass frequencies and attempts to squash everything down.

Additionally, there is a common issue with high-pass filters that you must be aware of. Due to the way filters work, if you try to draw automation that moves the frequency from a high value down to a low value in a split-second, you will get an enormous low-frequency pop in the audio. This extra surge of low frequency triggers a sudden attenuation from your limiter, resulting in a huge momentary drop in volume. See the problematic automation in Figure 9.1.

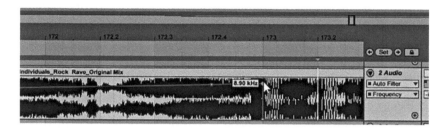

FIGURE 9.1
Sudden Jumps in High-Pass Filter Automation are Troublesome
Screenshot Appears Courtesy of Ableton

There are two ways to avoid this issue.

First, you can be smoother with your automation. If you need to get the high-pass frequency value back down in time for an important moment (such as the kick-in of the drop), start pulling the frequency down as early as an eighth note before the kick. See Figure 9.2.

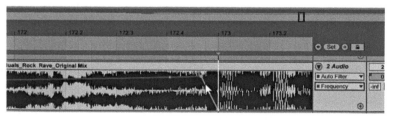

FIGURE 9.2
More Accommodating Automation
Screenshot Appears Courtesy of Ableton

Even better, if you have an element high-passed up, but you need to immediately stop the filtering altogether (i.e. to let it slam back in with full frequencies), simply leave the frequency value up high, but switch the high-pass filter Device off at the point when you need your bass frequencies back, via automation. By bypassing the Device, Live will perform a quick crossfade back to the non-filtered signal, free of pops. Then you can let the high-pass frequency settle back down to its lowest value whenever you like, because the effects of it are no longer being admitted into the final audio (Figure 9.3).

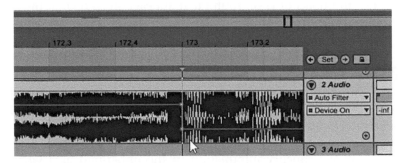

FIGURE 9.3
Bypassing the Filter at the End of the Sweep

Screenshot Appears Courtesy of Ableton

ADDED EFFECTS, SWEEPS AND DRUM BUILDS

Remember when we set up a folder for effects builds, whoosh/white noise effects and snare builds? Let's look at times that we should use them. Frequently, while filtering up out of a breakdown and gradually introducing the main track again, you will find that no matter what you do, you can't give it the energy and excitement it needs during the build. Sometimes it's because the main track contains a build sound that is appropriate in its own breakdown, but in the context of the new track, it doesn't sound energetic enough. Other times it's because you need to perform such subtle and gradual fades during the transition that the beginning of it has absolutely no energy.

In situations like these, we add our own effects. Try to find snare builds and effects builds that complement the tracks you're working with. Some effects have bright, electronic laser-beam qualities, others are smoothly filtered effects. Some snare builds are soft, sparse hits that don't go too crazy, while others are intense, over-compressed snares that perform rolls like machine-guns. Each has its place, and you want it to feel like the builds are part of the original elements, rather than noticeably sticking out. Some genres won't really welcome snare builds at all, and your energy must come from added up-sweep or whoosh sounds.

In sections where you need to abruptly cut off a track, or add drama to a particular moment, impact sounds such as reverberated booms or sine-wave sub effects can be handy. If you need to drop an element out and there is a sudden loss in energy, try one of these effects. This is particularly useful when coupled with reverb automation on the element you are cutting out. More on this later in the chapter.

Use these 'helper' sounds only if you need to. Less is more, and the less you need to add, the more natural the mashup will sound. The effects, drums and impact sounds the producer has picked out in the original piece of music will obviously be better matched to the song than what you pick. Occasionally you can even grab an impact or crash sound that appears on its own at the end of the track, and copy it over to an earlier section instead. This will save you trying to find your own sound, and will keep the added effects in context.

TREATING VOCALS

Sometimes, when using a vocal over an instrumental track, you may find that your vocals are hard to distinguish, even if they're turned up quite loud. Instrumental tracks are usually mixed to fill up all the frequency areas nicely, and often don't leave room for a vocal. The vocal will be 'masked' because there is already a lot of frequency information in the upper-mid area where the human voice lives. In this case, the simplest approach is to be sneaky, and use EQ automation to pull the clashing frequency band in the instrumental down, but only in the sections where you need your new vocal to play.

Loop and play back a section of the mashup where the two elements play together. On the instrumental track, either set up a new EQ or use an unused band on one of the existing ones. Push the gain of the band right up, set the Q or resonance to around 1 and play around with the frequency setting. Try to intentionally find the problem frequency where the instrumental seems to completely swallow up the vocal. If you're having trouble hearing it, search between 1 and 6 kHz. Though the frequency setting will affect the instrumental, focus your ears on the vocal instead. Once you've found the frequency where the vocal seems to disappear under the music, pull the EQ band's gain back below 0dB and listen closely to find out how far you need to pull it back until the vocal is clear enough. Try not to pull it down more than a few dB if possible, or there will be too noticeable a change in the instrumental. All we want is a subtle dip to help the vocal push out in front—pay attention to the moment when you 'see' the vocalist standing in front of the instruments. Note down how much reduction is required and set the gain back to 0dB. Finally, bring up the band's gain value in the automation lane. Throughout the arrangement, every time the vocal appears, pull the gain value down to the amount you have determined you need to. As the vocal ends, allow it to come back up. Try to keep it subtle and the vocal will be clearer and easier to hear without the change in the instrumental being too noticeable.

The other way to help a vocal blend in is to use the Return Track you set up in your template. On the right-hand pane of your arrangement window, each track has send settings, displayed just under the Track Volume (Figure 9.4).

Notice how we are adding reverb at the send control rather than doing it with automation. This is because we want the reverb send amount at a consistent volume through the mashup. See if adding a very small amount of reverb helps the vocal blend better with the music behind it.

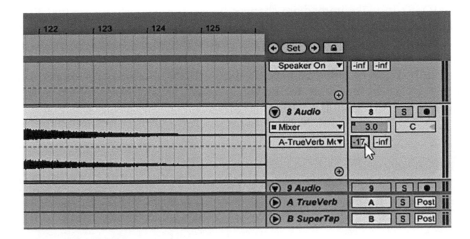

FIGURE 9.4
Sending Some Audio to a Return Track

Screenshot Appears Courtesy of Ableton

USING LOW-SHELF EQ TO BALANCE BASS

When setting up a Live template, we discussed how having the low-shelf band on an EQ is vital to avoid low frequency clutter and confusion on the dance floor. Let's look at the process of mashups and where to use this theory.

As a general rule, when two tracks that contain bass information are playing together, such as a main track and a break track, each track will be mixed and produced with its ideal amount of low frequency already in there. The kick will be nice and loud, driving the beat, and the bass-line and low-end instruments will take up as much space as they need to without making everything sound like a subwoofer party. Here's an analogy that might help (if you play DJ sets live, this is the same idea as mixing between tracks).

If you think of a single track as containing 100 percent bass information when playing on its own, consider that when you mix two tracks together, you still want the total bass information coming out of the speakers to equal 100 percent. So, either one track should have the bass pulled out completely, or each track sits at 50 percent bass. Through experimenting, you'll notice that the 'feeling' and illusion of the music will be dominated by whichever track is permitted the most bass presence. For pairs that contain different root notes, scales, energy levels, emotions or rhythm patterns, the track with the larger percentage of bass will ultimately decide which is *felt* more on the dance floor. People don't want jumpy transitions, but they are also used to hearing intentional changes in energy every eight bars, so when you automate the low-shelf EQ band, try to use a mix of gradual gain changes and sudden gain changes. Use the draw tool (CTRL+B or CMD+B) to make immediate changes between 8-bar sections so that they make musical sense, as new sections are introduced or concluded. See the below breakdown example, where the presence is shifted from the bottom element to the top element a little bit more every eight bars, by changing the bass balance between them. Remember, to have the bass at '100%' means 0dB of gain on the EQ, so it will appear as 0.5 on the automation lane (the middle) rather than 1.0 (as much gain as possible) (Figure 9.5).

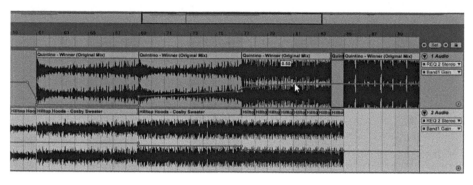

FIGURE 9.5
A Mix of Smooth and Sudden Bass Transfer During a Transition

Screenshot Appears Courtesy of Ableton

With this in mind, remember that transitions in or out of breaks must be treated with this bass reduction, as must hinting loop sections (explained shortly) played over a main track. Even if you are already using a high-pass filter to gradually shift the bass in or out of an element, it's still a good idea to use your low-shelf during these transitions to make sure you're not overloading the mashup with too much bass.

REVERSE REVERBS

Reverse reverbs are a very handy trick to have available when you need to bring something in 'cold', without the ear hearing any part of it beforehand. Let's take the most common example, a vocal. The mashup begins with an intro then a breakdown, before bringing in a vocal from a completely different track with no warning. Even if it is key and tempo-matched, the listener may get confused by the sudden appearance of something that sounds foreign. In order to help them prepare for the sound, it's a great idea to grab a very short section of the audio they are about to hear and use it to create a 'reverse reverb'.

Though a normal reverb is an imitation of the reverberation that we hear when a sound occurs in an acoustic space, we can create an interesting effect by sampling some reverb and reversing it. Though it has an unnatural feel, it does do a really good job of telling the ear that a new sound is about to come in.

It even gives it a heads up on the specific pitch and frequency information of the incoming sound. This effect may only last a second or two but it is enough to prime the listener for the element.

Begin by selecting an important piece of the sound. For a main track you are about to introduce, the best piece to sample is the most upfront element, such as a bass or lead note; usually the first note (try to sample the piece just after the kick drum so that the reverb isn't muddy). For a vocal, it's the first word. The sample shouldn't be any longer than a single beat. For an element that has drums in it, try and use a piece of audio that doesn't have a kick drum on it, even if that means using just the second half of a synth note (Figure 9.6).

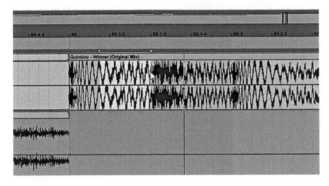

FIGURE 9.6
Creating a Reverse Reverb from a Selection
Screenshot Appears Courtesy of Ableton

Copy the selection to the end of the Live session (after your mashup ends) and move it down to an unused Audio Track. Then place a long reverb on the Device Chain for the track. Use a nice long reverb that is at least as long as you want your reverse sound to be. For convenience, you can even copy the reverb Device from your longest reverb Return Track. After the reverb, move your second EQ or filter (whichever plug you use for high-pass filtering) to the end of the chain, after the reverb. Making sure the filter is switched on, cut off any boggy low end mud using the high-pass frequency value, particularly the remnants of any kick drums. At the same time you want to leave enough low/mid in there to give the ear the heads-up it needs. Somewhere from 250–300Hz should do it, but listen carefully to be sure. Aim for warmth without muddiness.

Once you are happy with the tone of it, select a long enough section of time in the session to catch all of the reverb tail and export it, calling it something that makes sense to you, such as 'Reverse synth stab.wav' (Export is CTRL+SHIFT+R on Win, CMD+SHIFT+R on Mac). Remember that to export audio, it doesn't matter if the clip itself is selected or not; Live will simply pay attention to the *time selection* (see Figure 9.7).

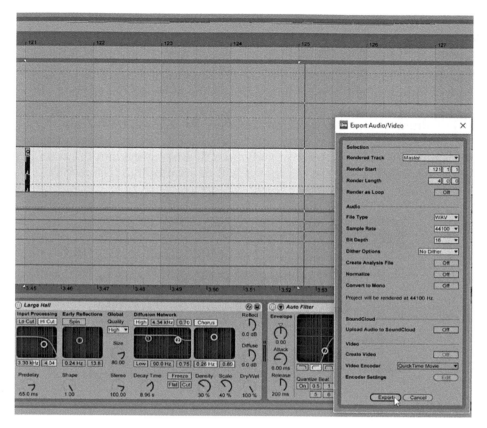

FIGURE 9.7
Export a Time Selection Long Enough to Include the Reverb Tail

Screenshot Appears Courtesy of Ableton

A 16-bit 44,100 wav will be fine. Export it to the same folder as your mashup. Afterwards, re-import the wav file onto a new Audio Track and access its clip properties. Click 'Rev.' to reverse the sample. You should also un-check 'Warp', as there is no point in time-stretching an effect like this. Adjust the clip boundaries if un-checking Warp has shortened or looped the clip. You will likely need to turn the volume of the clip up as well.

Move it to the perfect spot in the arrangement so that it ends just as the element begins. Listen closely to make sure the loudest peak of reverb doesn't occur too early. If it does, push the clip along until it ends right on the beat. When you play it through, it should sound like the reverb leads right into the word or note.

Much of the time, the listener only needs to hear a bar or two of reverse reverb to adequately prepare them. If you want to shorten the amount, use volume automation until it becomes audible later on the timeline (Figure 9.8).

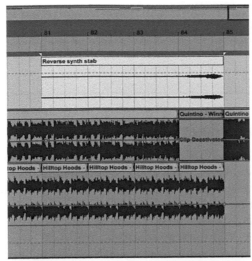

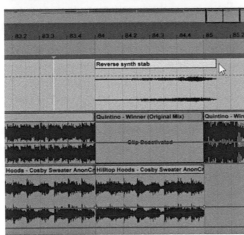

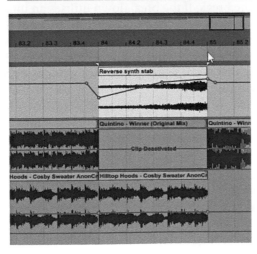

FIGURE 9.8
Making the Reverse Reverb Fit
into the Arrangement

Screenshot Appears Courtesy of
Ableton

> Once you become familiar with creating reverse reverbs, how long to set the decay times and where to set the high-pass settings, open up your template and set up a dedicated Audio Track up for creating reverse reverbs. This will save you a lot of time and effort in the future.

'HINTING' AND 'LINKING' SECTIONS

Often you will need to create a better blend between the different *sections* of your mashup. A great way to 'get some of the feeling' from your main track into your break (or vice versa) is to use subtle loops from one element under another. A listener may not be able to consciously notice why, but they will be able to feel more unification and harmony across the whole mashup. This is one of the most powerful tools you can use. You don't need to do this every time, just in instances where the feeling/emotion/energy between your break and main track are a little too different. It also provides you opportunities to add your own pickups, or add extra energy in sections that need it.

To use a piece of the main track over a breakdown, steal a loop from the intro or outro. I suggest you find one single bar, from the simplest section of audio you can, either in the first or last eight bars of the song. Try to find one that doesn't have any effects whooshes on it, or your ear will notice it looping; grab a bit from the middle of an 8-bar section. Make sure you put it on a *different* Audio Track in the arrangement, as you will need to apply effects and mix settings that are very different to your treatment of the main element. Use the Track Volume fader on the Audio Track to turn it down to at least −6dB, and use the EQ to pull the bass down as far as it goes. This will make the loop sound distant and subtle, and ensures there is no kick drum coming through. Now you can arrange it in your break. Use low-pass filtering or volume automation to creep it in slowly if necessary, and use volume automation and high-pass filtering to pull it out towards the end of the break. If the breakdown utilises an intentionally sudden change in energy as it goes into the build section, try ending the subtle loop abruptly to fit the breakdown. The important thing here is to keep the loop very subtle; it's there to support, not to lead. If you're not sure where to bring it in, try placing it at a spot that needs a little bit of pickup, or as a vocal chorus comes in. Check out the example in Figure 9.9. In order to keep the 'feeling' of the main track going in the break, it actually uses two loops. The first is a breakdown loop to keep the vibe of the synth going and make the breakdown feel connected to it. The second is a drum loop from the intro, used as a pickup and to prepare the listener to return to the main track. Without these two loops, the elements are just a little too different to feel cohesive.

Similarly, you can steal a loop from your breakdown track to use over the main track. It can be used to either prepare the listener for a breakdown, or to help build to it and tell the ear that something new is coming. Additionally, it can be used after the break to help bring the final section home. Again, use it subtly.

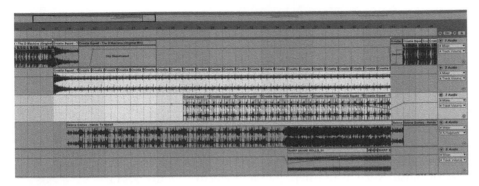

FIGURE 9.9
Using Loops from the Main Track to Add Energy to the Break

Screenshot Appears Courtesy of Ableton

Often it is necessary to use **side-chain compression** or an **LFO** tool to make sure it doesn't interfere with the kick drum, but only if the kick drum plays a regular four-on-the-floor rhythm (more on side-chain compression/LFO tool shortly).

Figure 9.10 is an example of a breakdown section being 'hinted at' over the drop. The audio in this section of the break element is already looping, so no manual loops were required. Notice the use of volume and low-shelf EQ automation to subtly bring it in over the space of eight bars.

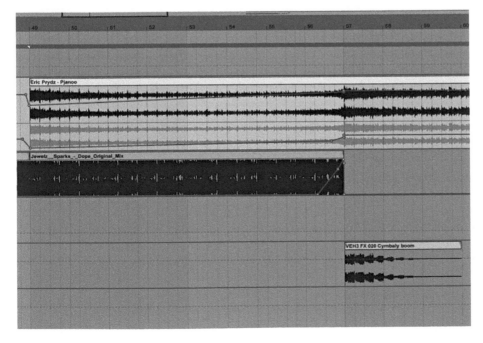

FIGURE 9.10
The Breakdown Track Being Hinted at During the Drop

Screenshot Appears Courtesy of Ableton

Following on from this idea, it's useful to know that you can help your break and main sections to connect better by using short sections of the breakdown track in your main sections as a feature, and to remind the listener of the context of the mashup as a whole. For example, halfway through a beats section, towards the end of an 8-bar block, you can 'drop out' (temporarily mute) the main track and play a bar or two of the breakdown melody. The theory behind this is the same as what producers do within their tracks: dropping in little snippets of the breakdown in their main sections to help tie the whole song together for the listener. Such tricks are much bolder than the subtle art of hinting and sneaking audio underneath, so they are great to use in very choppy, aggressive styles of music, such as dubstep, trap and electro (Figure 9.11).

FIGURE 9.11
A Bold Cut Across to a Single Beat of the Breakdown Song

Screenshot Appears Courtesy of Ableton

VOCAL LOOPS AND CHOPS

Looping and hinting don't just have to apply to full-bandwidth elements. Vocal loops and chops are just as useful for both preparing the listener for an upcoming vocal and for having the feeling of a vocal carry through to a final section that you don't want full lyrics in.

On a separate Audio Track with the same EQ and reduced volume guidelines in the 'hinting' technique above, pick a small section of vocal to loop. You can even use a section as short as one beat or half a beat. Try to pick a piece of audio from a long, sung note and create a clean loop out of it. Try to also use a note that feels comfortable over the music behind it. Again, you should consider using side-chaining if it seems to fight against the other elements when used over drum sections. Just like the idea we talked about when using 'hinting' loops over another section, the important thing is to keep the vocal loop subtle, never pulling more attention than the main elements. A common usage of this method is to 'hold' the last note of a vocal chorus, so that the note can extend all the way up to the end of a breakdown. To hear an example of what this sounds like, listen to the way the final note of the chorus holds in Fatboy Slim— 'Praise You'. If you want to get tricky, you can really chop up the vocal to make your own sentences, or fit the chops around the instruments in musical ways. Figure 9.12 shows a more extreme example where pieces of vocal have been chopped and arranged to create a repeating pattern, sitting in a rhythm that supports the main track. This pretty much makes it act like its own instrument.

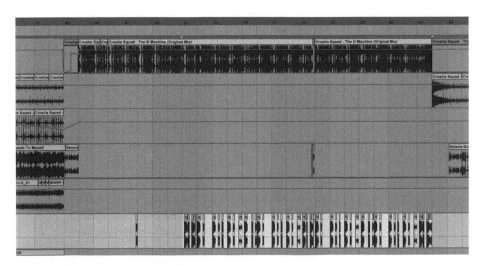

FIGURE 9.12
Vocals Chopped into a Rhythmical Pattern

Screenshot Appears Courtesy of Ableton

SIDE-CHAIN COMPRESSION (OR LFO TOOL)

When looking at hinting loops, I mentioned the use of 'side-chain compression' and 'LFO tools' to keep breakdown and vocal loops under control during your beats sections. Let's look at this in a bit more detail. Ordinarily, a side-chain compressor is an effect used in mixing where an incoming peak in audio from one source triggers a temporary drop in volume on another track. It's the kind of thing you'd hear on TV, where backing music is playing but it automatically drops in volume every time somebody on screen speaks. In dance music, the trigger to cause a drop is usually the kick drum, and the volume drop is applied to everything else in the mix. Extreme versions of this effect have become popular in dance music; for two examples, listen to the way the instruments 'duck' under the kick drum in the tracks 'Call On Me' by Eric Prydz and 'Satisfaction' by Benny Benassi.

Though this has become a very popular creative method for creating the 'character' in dance music, the technical purpose of this effect is to create sonic room for the kick drum, assist with the rhythm of the track and to keep the mix under dynamic control. In the mashup world, we don't have the main track's isolated kick-drum to use as our side-chain compression trigger, so if we want to create a ducking effect, we need a fake one. Therefore, rather than teaching you about how to use compressors, I'll show you how to create a fake side-chain effect using a low frequency oscillator (LFO) tool.

> Note: The LFO tool technique will only work with regular 'four-on-the-floor' rhythms. For more complex rhythms, it is better simply to go without.

Unfortunately Live doesn't have an LFO tool built in (at the time of writing anyway), but there are some great third-party alternatives.

If you have access to the *Max For Live* database, you can search 'LFO tool effects' for free options. The one I recommend, though it is not a free plug-in, is a VST plugin—*Xfer's 'LFO Tool'*.

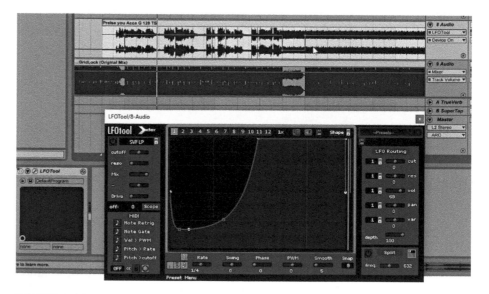

FIGURE 9.13
Xfer's LFO Tool

Screenshot Appears Courtesy of Ableton and Xfer Records

Low frequency oscillators were first conceived in early **synthesisers** as a way of creating subtle variation to synth sounds. Whilst synth notes are made up of very high speed oscillations, low frequency oscillators create very slow moving wave-shapes, and can be assigned to impose subtle undulations on volume, pitch or filter frequency levels to create subtle variance on the notes. The idea behind this plugin is to take that theory and apply it at a specific rate that syncs with the tempo of a project, such as repetitions of a ¼ note, ¹⁄₁₆ note, or bar. It can affect an incoming audio signal with an endlessly repeating pattern that applies change to volume, pan or filter frequency—based on the BPM of our session. We will set it to apply an effect that mimics what we would hear from a side-chain compressor reacting to a kick drum. The difference here is that the LFO tool will perform it in reaction to time, by applying a volume envelope to the audio signal on each beat. This way, we don't require a solo kick drum track as a trigger.

A good starting point is the preset, 'Sidechain-5' (see Figure 9.13). The preset has a side-chain shape written into pattern '1'. At the beginning of each ¼ note (determined by the *rate* at the bottom) the pattern loops, creating a quick drop followed by a slower rise back up to 100 percent. You can see the shape of this on the graph. In the LFO Routing section on the right, you can see that pattern 1 is being instructed to impose that shape on the volume of the signal, by an amount of 69 percent (set by the 'vol' slider).

Play around with the shape and depth of the envelope until you feel like it creates a rhythmic 'ducking' effect that sounds right for the track. The more you can alter the shape to feel 'in-rhythm', the more it will contribute to keeping the illusion of rhythm alive. If you're unsure, stick with the preset shape. It only has to be a subtle dip in volume, just enough to pull the sound out of the spotlight. You can reduce the amount of effect it has on the volume by bringing the 'vol' slider closer to 0. Another great bonus of using a tempo-locked plugin like this is that it will follow tempo changes in your mashup.

Don't forget to turn the LFO plugin off when you don't need it, for instance in your breakdowns. Otherwise, the volume will jump all over the place when it's not supposed to. Use automation to control the LFO tool's Device On/Off setting and only activate it during sections where the main track's kick drum is playing.

ADDED DRUM LOOPS

Sometimes, when combining two tracks for a mashup, you may find that there are certain 8-bar sections that lack a little energy, particularly during a breakdown that is a lot lower in energy than the main track, or during the build-up transition between breakdown and main track. You might also find that you need to create a point of difference between two breakdowns in one mashup that feel too similar.

Using added drum loops during breakdowns and build-up transitions are great for these situations. Using a simple loop with a neutral rhythm can add subtle energy lift to a section without affecting its key or emotion.

In both situations, ensure that you listen carefully to the tracks underneath and pick drum sounds that match the style of the track you are adding to. This will help them feel like they are part of the track instead of foreign elements.

First, let's look at breakdowns. Many breaks taken from radio or classic dance tracks will be long ballad sections with little or no drums. This may be great when listening to the original song, but remember that in a mashup we need the energy level to come back up at least a little before we transition back to the main track, particularly if it is a big, powerful, high-energy one.

To help get us back to the peak energy, we can introduce one or more subtle drum loops during the break. The simplest rhythm is a drum hit on each beat of the bar. Be careful if your breakdown uses the complicated rhythms of breaks, drum and bass, or dubstep drums, because simple '4 on the floor' rhythms can work against them. In terms of which drums are likely to work, try the following:

- Kick: A simple kick drum helps add that thumping weight to the break, but will need to be silenced when the build-up begins. Experiment with volume and low-pass filtering. Too much kick presence makes the break feel like a main section, and confuses the listener. Listen to how kick is used in the breakdowns of tracks in your genre for guidance.
- Claps: Whilst claps used to traditionally appear only on the 2 and 4 beat of each bar, a clap on each beat is pretty common with party-style tracks, and also became very popular during big-room music from around 2012–2015, so dance floors are used to hearing them. They can really assist the energy in a section by adding the atmosphere of a crowd, as well as providing a solid rhythm element for the ear to anchor onto. Pay attention to the track underneath when choosing between 'real-sounding' handclaps and overtly electronic claps.
- Ride cymbal: A ride on each beat works wonders for providing intensity without affecting the rhythm or feeling of the track. It may need to be subtly faded out toward the end of the breakdown to leave some energy for the final drop.
- Fuller drum loops with high-hats and percussion: This should be treated as a last resort. These really do help with energy, but they have the highest chance of sounding foreign. This will affect the feeling of the music a lot. Rather than adding your own sample, it's safer to try and find hats or percussion by stealing a loop from one of the tracks, and treating it as you would a 'hinting loop'.

If you plan on using more than one drum loop, pay very special attention to *where* you use each one. Every track is different, and you want to make sure you introduce feelings of intensity (ride), warmth and power (kick), and crowd-like momentum (clap) in an order that suits what is happening underneath. Also pay attention to any vocals in the breakdown; new drum elements can distract from a vocal coming in at the same time, but can be placed cleverly to increase the power of a vocal as it moves into the chorus.

Similarly, the build-up is a good place to use a drum loop if you find that the transition section doesn't build up enough before the kick-in. Of course, as we have discussed, your first priority should be to make sure you have the necessary snare builds and build effects to help the transition work, as these are much more effective. However, if you still discover a need for more energy, try an added clap or ride loop. If the builds or track underneath fade or filter out toward the end of the breakdown, try to match the filtering using your own volume and filter automation on the drum loops. Listen carefully to make sure everything disappears at the same rate.

USING YOUR OWN INSTRUMENTS

For those readers who have experience writing and producing, there are a few instances where using your own instruments can be incredibly helpful, either to assist a transition or to add something to a section of the mashup.

When assisting a transition, an instrument can be used in the same way that a snare build, drum loop or white-noise effect would, helping to repair a drop in energy or support a weak build-up. These are a little trickier, as they not only need to be in the right key and rhythm, but they need to be instrument tones and melodies that fit with both tracks in the transition. The simplest instrument you can use is a 'high-string', named because this kind of melody line is traditionally performed with orchestral strings.

Using either a simple synth or an actual string patch, you can use a single high note bridging over the transition section, in the key of the mashup. This can help solidify the key of the mashup in the listener's head, as well as providing something common that the ear can hang on to while the rest of the mashup transforms from one song to the next. Whilst this certainly isn't something that you should need to use very often, it can be helpful during transitions where the key becomes hard to keep track of. In terms of specific synths to use for this, I often use a simple sawtooth patch from Lennar Digital's *Sylenth1*, or for a simple orchestral string patch I look to ReFX's *Nexus 2* or Native Instruments' *Kontakt 5*.

When working on a mashup that needs a little more emotion and power in its final 'kick-in' section, adding a subtle high-string can be helpful here too. This will work best if the main track element (drop section) is not too overcrowded or busy. The important thing here is that it *must be subtle*, or you'll likely overcrowd the music.

The most rewarding situation to use this is when you have a main track that is big and tough, but not particularly emotional, and during the breakdown you transition to a very emotional track. When you come out of the break and back into the main track, rather than let all of the emotion drop away, you can keep a little of it going by putting a high-string note in the key of the track over the final beats section. High-strings help carry some emotion through into the final section and give *new meaning* to the main track.

PITCH EFFECTS

A handy Device that comes bundled with Live is *Frequency Shifter*. It's a brilliant effect that can add a lot of excitement to a vocal loop that you have been playing towards the end of a breakdown. Often you either need that little bit of extra excitement as you filter your vocal up and out of the break, you need some help transitioning the vocal out of the spotlight, or you just need a special feature effect to make the second break more exciting than the first.

Though it can be used on anything, I find it works best on vocals.

To use it, place it in the Device Chain before the EQs. The setting you want to manipulate through automation is 'Frequency'. Leave the other settings at their defaults. Making sure the value remains at 50 percent (or 0Hz) for the rest of the track, have the pitch automation rise slowly over your desired time, usually an 8-bar section. Be careful not to let it rise too high as it can start to sound a little ridiculous when pushed to its maximum pitch. Since pitching up like this is usually a way to finish off a vocal section like this, it is usually best to use it alongside volume or high-pass filtering automation to gradually remove it from the mix.

This is a very clever Device, and doesn't present the delay time and CPU-load problems of a regular pitch-shifter, which would basically involve performing a constantly changing Warp calculation. Instead, as the frequencies move up, their relationship with each other falls apart, and the sound takes on a 'ringing', metallic sound that is great for transitioning it out when combined with high-pass filtering or fading. To hear this kind of effect in action, listen to the looping vocal in Armand Van Helden's remix of Ou Est Le Swimming Pool—'Dance The Way I Feel', as it frequency shifts during the breakdowns.

ARRANGING TEMPO CHANGES

So we've already discussed tempo changes in regards to how to import tracks that will require them. But how do we make them work in our arrangement? What extra techniques do we need to incorporate to ensure we perform a tempo change in a way that doesn't empty the dance floor?

If you've ever tried abruptly shifting the tempo up or down during a set before, you may notice that it completely disorients the dance floor. Dance music is all about that steady, predictable beat. The brain can handle a certain amount of correction, but shattering the illusion of the music by messing with the audience's expectation of the rhythm is the quickest way to get yourself in trouble. By a strange psychological effect, the weight this has when you hear it on headphones or in your studio is very small compared to the jarring feeling it creates in a nightclub. We can remedy this by being subtle and gradual in our tempo change, by giving the listener a rest from the beat, and using effects to mask the change.

First, let's revisit what we talked about in the earlier chapter regarding breakdown tempos. We used the example of an R&B tune that usually sits at 90 BPM, and tried to push it closer to 100 BPM for the mashup. At this point you should make sure you have decided your intended tempo for the breakdown, as this affects the rest of the following steps. Don't stress if you don't get it right the first time—you may find that once you add in all your other techniques you can allow the tempo change to be a little greater.

As you approach the breakdown where you intend to make a tempo change, start slowing the song's tempo very gradually. You only want to do this for around 1–4 bars. You don't want the tempo to actually reach the breakdown speed by the time the break starts; instead, let it get to about halfway there.

At the point where the main track drops out, you need to make a choice about how to start the break. For tempo changes, there's really two ways to do it. You can either continue straight into the new element, attempting to make the transition as smooth as possible, or you can use a safer technique, and include a 1–4 bar gap before you let the breakdown element start. Let's look into the safer option first.

Leaving a gap before the new, slower element gives the listener a chance for their brain's internal metronome to reset, and when the new element comes in at a slower speed, they are able to process it more easily, noticing the change much less. So long as you fill the gap with something, it will be ok. See Figure 9.14.

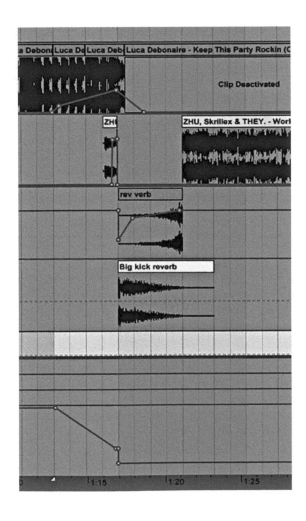

FIGURE 9.14
A Mashup That Slows Down for a Lower-Tempo Breakdown

Screenshot Appears Courtesy of Ableton

Obviously you can't just cut to dead silence for two bars or it's going to sound like a mistake. It's important to use some kind of effect at the beginning of the gap, as seen in the figure. Depending on the feel you want to go for, you can use a sample such as a white noise down-sweep or a boom. Alternatively, if you want to create the feeling of the main track dropping off into thin air, you may avoid the use of any impact sounds and simply have the main track automate to the long reverb send. I often use both of these ideas at once to create a big impact but still have the feel of the main track 'hang over' into the gap. Either way, you'll want to have *some* audio spilling over before the breakdown track drops in. The duration of your reverb or effect sample may affect how long a gap you want to leave before bringing in your breakdown element. Above all, go back, play it right through to see how it sounds—and go with what feels good.

This is also a good moment to try a reverse reverb (explained previously) to announce the incoming breakdown element, particularly if it is an acapella or a very famous song riff.

Never use delays over a tempo-change. Because the lengths of reverbs are based on a duration of time, and have nothing to do with rhythm, they mask a tempo change quite well. Delays, on the other hand, are effects we set up to intentionally hit in sync with the rhythm of the mashup. When used over a tempo change, the delay will do one of the following (depending on the plugin). The delay effect will either continue at the same speed over the tempo change, resulting in delays that hit completely out of time. The Device will respond to the tempo change and spread the audio samples out to fit the new tempo, and the delays that are already triggered will change pitch rapidly and sound like a record player losing power. Or, the plug-in will reset each time Live sends it a new song tempo value, meaning that during a tempo change section it will output sudden silences, partially cut-off delays and nasty glitches.

If you have decided to use the gap option, you will not need as great a tempo slow-down in the lead up. You may find that you don't need to gradually slow it at all, but just be careful. A subtle slow-down in the lead up tells the listener's brain to expect a tempo reset, which makes the upcoming breakdown easier to process.

If however you decided that it's *absolutely crucial* to have your main track flow directly into the breakdown element, that's ok. One possible reason for this is that you are using a vocal or other element over the top that needs to continue without interruption. Another reason is if you are worried that the gap will ruin the momentum of the mashup.

If you need to keep your elements going, give the first four bars of your breakdown some space to settle into the new tempo by continuing the gradual tempo change we used in the main track. See Figure 9.15.

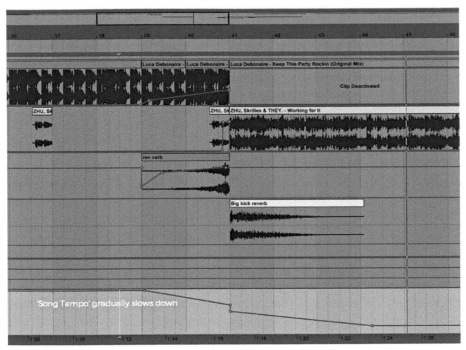

FIGURE 9.15

A Gradual Change in Tempo Prepares the Audience

Screenshot Appears Courtesy of Ableton

It's still important in this case to use samples and effects returns to help mask the tempo change, but be careful using booms and sub effects if your new break element contains low frequency information such as drums and bass-lines. Sending the main track to some reverb in the lead up to the break is fine too, just use your ears and maintain a good balance between using reverb to mask the tempo change and using little enough to avoid a total mess. A good white-noise down-sweep at the start of the break is a good way to smooth over the energy dip as all the elements of the main track disappear. Don't forget to turn Warp *off* for all booms and noises, otherwise they'll try to stretch as the tempo changes.

Be careful as you draw in your tempo automation. When performing a tempo-changing transition with no gap, listen very closely to the energy level between the two songs. Just because the BPM is slowing down, doesn't always mean that there is less energy. When applying your gradual tempo movement, use your ears more than you use the visual representation of the automation. Imagine you are on a dance floor as you hear it and decide how it would make you feel. An R&B track at 110 BPM can have more 'energy' than a house track at 120 BPM, even though the rate of beats is technically slower. Listen on speakers to maximise your chance of *feeling* the effect as the dance floor will.

As we discussed earlier in the chapter, you can try to use those subtle loops taken from the main track element during the break to assist the continuity and make the speed difference less noticeable. Even though the loop will be playing back at a slower tempo, it will still help to keep the listener's mind connected to the main element, and will make it easier for them to return to the main track at the end (if you are returning to the same song). Because Live has to stretch the beats out so far for this kind of time-stretch, any loop from your main track should have its Warp set to ¹⁄₁₆ Beat Interval during the break.

The best way to get back out of a down-tempo break and into your main track is to take advantage of the build section that pretty much all dance tracks have. Usually it's an 8-bar section (occasionally 4, 12 or 16), and you can use this moment to bring the speed back up. During this section, try to keep the breakdown track running over the top in one of two ways.

The first option is letting it run as normally, continuing as it was. Musically, it should be a section that gives the breakdown a strong finish such as a chorus section. If the build-up is too messy, usually due to chord changes in one of the elements clashing against the other, pick the other option: take a short sample of breakdown audio from 1–4 beats in length to loop during the build. The best places to sample this are usually either a single beat right before the build begins, or the very first beat or bar in the build section. Choose primarily with the music in mind, what makes sense within the whole mashup, and secondarily with smoothness in mind, since you want the audio to loop nicely.

To make your choice of loop a musical one, try to use a section or phrase that fits key-wise with the section coming in. This can be difficult, but some experimenting should help you feel when some loops *do* fit over the build and some loops *don't*. You might even find you have to pick a sample from the third or fifth bar rather than the first within an 8-bar section, depending on the chord progression in the break track.

Use the build section of the main track to decide how long the build section in your actual mashup will be. Before applying effects, automation or extra samples, first arrange the breakdown section/loop and main track to occur where you want them to. Click on the spot in the main track where the drums start playing again (the 'drop'), and make a splice in the clip (CTRL+E Win, CMD+E Mac).

Remember that when you have spliced a clip into two, you can change the Warp settings on one clip and it won't affect the other.

Though we have discussed it in the earlier chapter on Warping, let's refresh on the setting required for this tempo-changing build section. Select the clip that represents the main track's build, making sure Warp is enabled for the clip and that the markers are set up properly. Then, set the beat interval to ¹⁄₁₆. The slower tempo required near the beginning of the build will be too much for Live to stretch at ¼ intervals, and the ear will noticeably detect drum or instrument hits in the wrong place; ¹⁄₁₆ ensures that it can catch all of them. Of course, once the mashup exits the break and returns to kick-in section, be sure to go back to whatever Warp setting (or Warp-off setting) you had. (Make sure you repeat this process if slowing down *prior* to the breakdown as well, temporarily changing your main track's Warp resolution to ¹⁄₁₆.)

Next, set up the Song Tempo automation to get the mashup back up to the main tempo. Start the rise in automation as soon as the build begins, and have it reach the final tempo around two bars before the kick-in. It's important to give the listener a couple of bars at full speed to re-acquaint their brain with 'dance' tempo before they're required to actually dance (or fist-pump the air, depending on your crowd). This is super important! The audience is very sensitive about being hyped with 8–16 bars of anticipation, only to be deceived about when the beats will kick in. A tempo-changing mashup puts extra responsibility on you to ensure that the rhythm is *very clear*, or you will lose the illusion of rhythm. You might even have to make the beat more obvious by adding drum elements of your own.

Listen through the build closely. Make whatever changes to the slope of the tempo envelope you need to in order to make sure the change sounds as smooth as possible. Depending on what kind of drums, instruments or vocals are present during the rise, each build will require a slightly different approach. A straight line on the automation page doesn't always feel like a steady acceleration to the listener.

As with a non-tempo-changing build, make sure you pull out the break track before any drop-out occurs in the main track, such as a 1–2 bar gap for a vocal shout, drum fill or hook melody. Just ensure you've reached full speed by the time this occurs, or the audience will lose the rhythm (Figure 9.16).

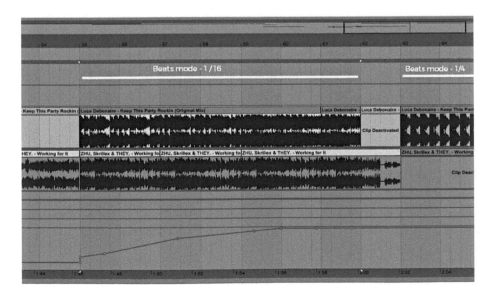

FIGURE 9.16
Accelerating Back up to Dance Tempo

Screenshot Appears Courtesy of Ableton

Once you are happy with your tempo transition, then you can use any samples, effects Devices, automation, or filtering you think are necessary (remember not to use delays until after the tempo has returned to a static value).

If your breakdown contained a vocal (whether an acapella or as part of the breakdown element), often a great way to get back into the main track is to give the audience one more vocal hook to tie it all together before you kick back in with the main track. Keep in mind that though the vocal will play back at a tempo faster than normal, the brain is super excited for the beats to kick back in, and in this moment is more forgiving of speedy vocals so long as they are kept short (1–2 bars maximum).

Figure 9.17 shows an example that slows down into a break and speeds up into the drop again, using the techniques above. Notice the 'Song Tempo' automation down the bottom.

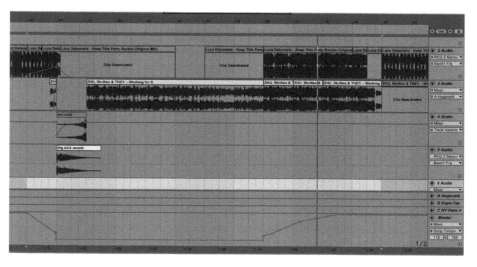

FIGURE 9.17
A Mashup with a Slow-Down Breakdown

Screenshot Appears Courtesy of Ableton

TRANSITIONING BETWEEN 4/4 AND 6/8 TIME SIGNATURES

When performing transitions between triplet and 4/4 time records, you must be careful how they overlap. If you test the two competing elements together and they work without clashing rhythmically, that's fantastic. Often though, they won't, and the duplet and triplet subdivisions will hit off-time from each other, causing a clash in rhythm. Sometimes, due to sparsely placed snare hits at the beginning of the build, the triplet rhythm won't become obvious until the last four bars, meaning you can make a gradual transition from one element to the other using volume and high-pass filter automation, and get away with little or no obvious rhythm clashes.

If there is still clashing however, the most successful way to make it work is to make the whole transition section either 4/4 or 6/8 time, rather than a mix of both. This means giving most or all of the volume and bass preference to one of the elements (most likely the breakdown element), and adding your own elements to help fill in any missing energy: snare builds and white-noise effects.

If you are moving from a 4/4 break into triplet time, you can usually get away with having the build-up remain in 4/4 time, but having it slam straight into triplets as it comes in with the beats. To help the transition, this means you generally need to use a reverse reverb made up of the main track's audio to help prepare the ear for it sonically.

USING MULTIBAND COMPRESSION TO MAKE ROOM FOR VOCALS

> Another thing I learned from a friend named Justin Blau, a.k.a. 3LAU, is a way to better fit vocals on top of drop sections. He once showed me a really simple and effective technique by using multiband compression side-chaining the high end of the instrumental using the vocal channel. So for any non-producers, whenever the vocal is playing the higher frequencies will be lowered and come back up whenever the vocal goes away. It helps because it allows more room for the vocal without compromising the low end (kick, sub etc).
>
> (Pierce Fulton, DJ/Producer)

If you feel like trying a more advanced technique, this helper can provide a little bit of room for your vocals when working over a track that is a little too full in the mid-range. If you find that there is an instrument such as a big, honky synth or a treble-ish piano playing, it might mask the vocal and prevent you from understanding the lyrics. Turning the vocal up or the track down may sound too unbalanced, and as a result it's necessary to pull out a little of the masking frequencies from the instrumental. The technique I showed you earlier on treating vocals—where you temporarily pull down an EQ band on the music behind the vocal—is perfectly fine for this, but if you are using an acapella, a multiband compressor can do this for you automatically. The multiband compressor only takes action when the acapella actually produces a signal. Therefore, it may be a good idea to try it if you have an acapella that plays so much of the time that it would be time-consuming to automate around.

As Pierce explains in the quote above, you can help give an acapella some space by setting up a multiband compressor over the instrumental. Set it to side-chain and select the vocal track as the key/trigger input. If you set it up to only affect a band of frequencies around the vocal, the instruments and mid/top-end information will duck out of the way to make way for the vocal, leaving the foundation elements of bass and kick to continue playing at full power. Compression is a very tricky beast, and I won't go into teaching you how to use it in-depth because it generally isn't needed in mashups, beyond what is covered in the book. I will however help you set a basic vocal–instrumental ducking compressor using Ableton's *Multiband Dynamics* Device, available in Live Standard or Suite.

Drag the Device from the explorer into the Device Chain on the instrumental track. Place it before the EQs. The Device works by splitting the incoming signal into three frequency bands: low, medium and high. It then applies separate compression to each of them and sums them together at the output of the Device (Figure 9.18).

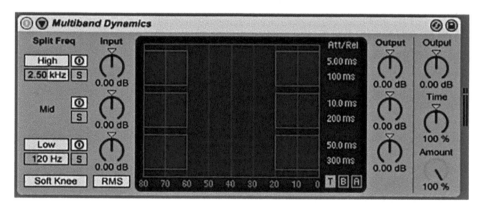

FIGURE 9.18
Live's Multiband Dynamics Device

Screenshot Appears Courtesy of Ableton

Begin by clicking the small power buttons next to the high and low bands, to bypass the top and bottom band. (Bypassing these bands will allow them to pass through unchanged.) Set the bottom crossover frequency to 600Hz and the top frequency to 5kHz. At the bottom-right, click the 'A' to show the 'above' settings, which allow you to set the volume threshold *above* which compression will begin. Set the dB value to −15dB (the threshold) and the ratio to 1:2. At the top-left of the Device, click the down arrow to show the side-chain section. Turn side-chain on, and set the track to the Audio Track the acapella is on. Turn the gain just under it down to the lowest setting (Figure 9.19).

Now, make a loop and play a section with the vocal and instrumental playing at once, where they seem to mask each other a lot. Slowly turn up that gain knob in the side-chain section until you hear it start to have an effect on the instrumental. You can also see when the compressor starts to kick in because of the orange numbers in the graph. These indicate how much the middle band

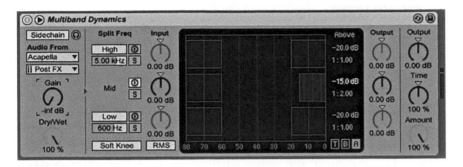

FIGURE 9.19
Multiband Dynamics Set up for Side-Chain Compression on the Middle Band

Screenshot Appears Courtesy of Ableton

is being attenuated in dB. Use your ears carefully and be very subtle. Remember that if a bunch of frequencies are sucked down to silence every time the vocal plays, it's going to sound unnatural. Find a balance between vocal intelligibility and instrumental preservation. If you pull your whole instrumental back a touch, you might also find you don't need quite as much side-chain effect. If you feel confident, feel free to play around with the compression ratios and crossover frequencies too.

GENERAL IDEAS ON WORKING EFFECTIVELY

Lastly, some very general things to keep in mind when working on mashups. These are simple principles that apply to any creative effort involving crafting music on the computer.

- If you start to feel doubt about your mashup idea or how you have arranged it, take a 5-minute break, or work on a different mashup for a while and come back later. Once the memory of how the mashup currently sounds fades from your mind a little, you will be able to hear it fresh, making it feel like you are hearing it for the first time. This gives you clarity and helps you determine whether general ideas or specific moments actually work or not.
- If you simply can't make a mashup work, it might be a sign that it isn't meant to be. If you need to employ ten of the strategies listed above just to make one transition work, the two elements probably don't belong. Things should feel natural, and the less work you need to do to make it sound good, the more confident you can be in the original idea.
- Err on the side of being more spacious, at least to begin with. People will react better to longer music that works than they will to fast-paced ideas that are confusing.
- Check over every change you make by rewinding a couple of bars and listening through to see whether it had the desired effect. If you don't listen to your changes in context, how can you know if they're right?

IS IT FINISHED?

Once you believe you have achieved what you set out to do, it's time to test it.

Grab a piece of paper and a pen. Rewind to the start of the mashup, hit play and step away from the computer. If you can, listen to it on your speakers, rather than headphones. Make sure you listen right from the start to the end. For this test, you need to change your mind-set. Instead of listening for creative ideas to add, you need to stop being the producer and become the audience.

So, imagine you are on a dance floor as you listen. Even close your eyes and visualise it. Feel the music and notice whether anything jumps out at you as too abrupt or confusing. Does anything pull you out of the moment? It is important to look away from the screen while you listen. Without a visual warning that new elements are about to come in, you are forced to use only your ears, and you will be able to hear things happen exactly as your future audience will. Believe it or not, this change in mind-set makes a big difference to how you hear things, and sometimes you realise you have missed the most obvious problems.

If you find changes you need to make, *don't* stop the music, write them down and keep listening. Afterwards, change everything one at a time, then repeat the test.

If you are happy with it, check through once more, but listen on headphones. This time, all you're listening for is clicks, glitches and pops that can occur while editing, usually where you have created splices in audio. It also helps you catch Warp settings that might not be 100 percent right. It's important to do this check on headphones because these glitches are often hard to hear on speakers. Once you are sure the edits are clean, it's time to export!

EXPORTING

So, after careful work and a focused listen from start to finish, you have decided you have finally finished your mashup. What now?

It's time to export the mashup into a file you can take with you to your DJ sets, so that everyone else can hear how good it is.

Do the following (in order):

1. Make a selection on the Arrangement page to tell Live what you want exported. Remember that in your session, the mashup might not necessarily start right on 1.1.1, or end at the last audio clip. You don't want an extra minute of silence sitting at the end of your file, or any leftover pieces of audio files you kept at the end of your timeline. More importantly, you don't want to select the wrong sections and accidentally cut off the start or end of the track (yes, I've regrettably done this a few times). The easiest way to make sure you get what you want is to make a selection using the *first* and *last* piece of audio in the mashup. Zoom right out and click the

very first piece of audio, which should be your intro. Then, holding shift, click your very last piece of audio, which should be your outro. Don't worry about which *tracks* or *clips* happen to be included in that selection, Live will only be paying attention to the *time selection* (Figure 9.20).

FIGURE 9.20
Creating a Time Selection for Exporting

Screenshot Appears Courtesy of Ableton

2. Go to File and click *Export Audio/Video*. In the dialogue that opens up, make sure the following settings are selected. Most of them will be already set by default, but check anyway (Figure 9.21).

    (a) Rendered Track: Master
    (b) Render as loop: Off
    (c) File Type: WAV
    (d) Sample Rate: 44100
    (e) Bit Depth: 16
    (f) Dither Options: No Dither
    (g) Convert to Mono: Off

Make sure you export to the folder your Live session is in. Live automatically directs you to the folder it last exported a file to, which usually happens to be the *last mashup* you worked on. So just be careful.

Also make sure you name your file properly. Label all the necessary information in whatever order works for you, but stick to the same format with every mashup. Make sure you include the key. I would suggest something like 'Title (Your name's Mashup) KEY.wav'—for example:

'The Greatest Mashup Ever (DJ Awesome Mashup) F#.wav'

If you like to have 'Mixed In Key' information in the title, that's fine too, just try to leave the musical key in there to make things easier for you later on when you come back looking for more mashup ideas.

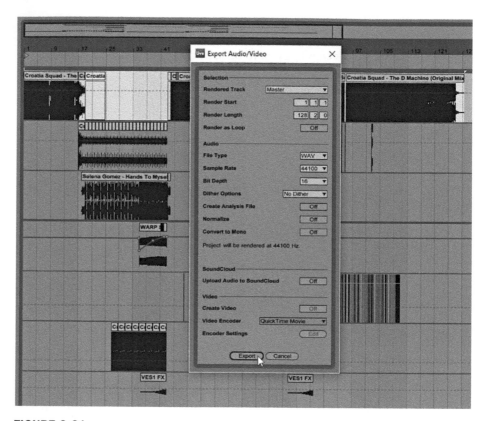

FIGURE 9.21
The Export Window

Screenshot Appears Courtesy of Ableton

Also, if it's important for you to have the artists' names on the file title, put that in too, but keep in mind your file name will start to get pretty long, and get cut off on some file systems or DJ hardware.

Note: In future versions, Live may include mp3 or other audio formats as options for exporting. If so, feel free to export in these types, so long as they are universal. It will remove the necessity for the following step.

3. Take your finished wav file and convert it to mp3, if that's how you like to keep your DJ music. Of course, there is always an ongoing debate about the fidelity of using CD quality wav versus mp3s, but rather than influence you one way or the other, I'll leave that decision to you (and I'll stay out of the debate!). Just make sure the mp3s are 320kbps (kilobytes per second). Any lower than that will sound awful when played out. To convert to mp3, I would recommend a fast, batch-converting application. On Windows I use a very useful and free program called Flic Flac. If you leave the program on a 320kbps mp3 setting, you can simply select one or more wavs in your explorer and drop them onto the Flic Flac window, which then creates mp3

duplicates. There are many more out there though; even iTunes can be used for it, although it is not quite as quick and convenient, and leaves both versions hanging around in your iTunes library. After it is converted, you might want to include the various artists included in your mashup in the mp3's *artist* tag properties, if you didn't add them into the file name. For example: 'DJ Legendary Guy vs 90s Classic Group'. Up to you, it's a bit of extra effort but it makes searching for mashups easier when you can't remember the name you gave it, which can be helpful at gigs. It's also respectful to the original artists to keep a record of whose music you're using, in case anybody asks about them after hearing the mashup.

4. Copy your mp3 to your 'Finished Mashups' folder. Remember when we set this up earlier? Leave a copy of the mp3 in the original session folder too—just as a backup.

5. Import the mp3 from the *Finished Mashups* folder, *not* the session folder, into your DJ music collection in Rekordbox, iTunes, Traktor or whatever you use. Remember, you're now going to move the session folder into an 'old' folder to get it out of the way—since it's finished. We don't want your DJ music program to lose track of your new mashup. How you organise it into your sets from here is totally up to you!

If you didn't check out the videos that accompany this chapter, remember they are there to help you see and hear these techniques performed, which can be quite helpful. You can follow along with Live open and try the techniques yourself!

House-keeping

Hopefully by now you have had a go at your first mashup using the guidelines in the previous chapters! If not, it would be a great idea to go and make your first batch now, and re-read individual techniques each time you use one for the first time. As you make more mashups you will get better and faster at picking the right techniques to help you solve different problems. You'll likely come up with your own techniques and shortcuts to getting your elements to fit together too. Now I'd like to share some ideas on how you can get even faster and more efficient as you continue to build up a collection of successful mashup ideas.

Keeping organised may seem annoying now, but trust me, when you've got a gig on the next day and you need to come up with some new material, being organised is going to help you create a batch of good quality mashups incredibly fast.

BECOMING FASTER AND MORE EFFICIENT

Watching my mate Andy J using a template he created in Ableton was great. He had every kind of chop up effect, EQ, compressor, etc. ready to go. Once the tracks are chosen, a good mashup should take no more than 15 minutes MAX to do.

(Ivan Gough, DJ/Producer)

Doing a lot of mashups allows you to have more in your repertoire than everyone else. It's going to help you get more gigs. I have to play so many different styles of music, and so many different venues that I need to have a very high output.

(Phil Ross, Commercial DJ)

The whole idea behind being as organised as you can is to cut down the time it takes to come up with a group of great ideas, to ensure that you already have everything you need when you need it, and finally to make it easier for you to produce them in a very short space of time. The quicker you work, the more confident your choices of technique will be, and it will keep you from getting lost in what you are working on. It also increases your output, and therefore your musical options at gigs, making you a more flexible DJ.

QUICK RE-CAP ON ORGANISATION

Let's pretend for now that you have just finished a batch of mashups. Here's what you can do to stay organised and make things faster in the future.

1.   Clear your Mashup Sessions folder
2.   Update your Mashup Keys list
3.   Clear your Current Mashup Ideas list
4.   Update your Classics/Musical Hooks folder.

First, let's go over the folder system you set up in the chapter about preparing for mashups.

Each time you finish and export a mashup, make sure Ableton Live is closed and move the session folder for that particular mashup into your 'old' subfolder. At first this may seem a little pointless, but once you have a collection of 20 finished mashups, it's going to be impossible for you to remember which ones you are working on and which ones are finished. If every finished mashup is neatly hidden away in your 'old' folder, you won't get confused and misplace anything.

After finishing a batch of mashups, figure out if there are any elements you want to keep track of for future mashups. You may have used a particular classic song for the first time, or a new dance track that you believe will be a big hit for a long time, and want it to appear on your Mashup Key list next time you're looking for ideas. Go ahead and update your list. Just remember to list your vocals in the 'Acapella' section and the musical ideas in the 'Riffs/Breaks' section to make sure you are reminded about the right elements at the right time. You'll find that if you make sure you add every relevant song into this Mashup Key list that your list will build up very quickly, and soon you'll have so many ideas to choose from that you'll never miss a great mashup opportunity.

Next, clear your Current Mashup Ideas list of all the ideas you have finished or no longer plan to work on. You could also think of this as your to-do list. As well as clearing the list, you clear your brain by knowing there's nothing left waiting for you to do. This is important, as you need to be able to depend on this list to accurately keep track of your many mashup ideas, and by their very nature, too many mashup ideas are hard to hold in the imagination. Keeping this list neat and current allows you to depend on it more, keeping your mind free to focus on one project at a time.

Lastly, update your Classics/Musical Hooks folder. Add in whatever you feel will be useful in the future. As well as just keeping full copies of old classic tracks, I often export the specific part I used in a mashup and put it in the folder, because I know that's the exact part I'm going to use next time. Sometimes

I have a classic riff that I had to do a lot of careful editing or Warping on to get it to fit into a dance music tempo, and I'll export this so that I don't have to go through the process again next time. *Put in anything that will save you time in the future.*

USING YOUR OLD MASHUPS AS INSPIRATION

During the stage where you are creating your mashup ideas list by matching together songs by key, try also looking at your finished mashups for inspiration. If you are looking at your Mashup Keys list and feeling like none of the vocal ideas are special enough to use, go into your music collection and search for mashups in the right key. You'll often find good ideas that don't necessarily appear in your Mashup Key list. This is one of the reasons it is important to put key information in the title of your finished mashup. Sometimes this will be incredibly useful because a single mashup may have more than one element you'd like to use. This brings us to the next topic.

COMING BACK TO YOUR PREVIOUS IDEAS

At some point you are going to want to re-use a classic breakdown or a featured acapella in a new mashup idea. The above house-keeping tips create a great foundation for you here, allowing you to take your pre-edited elements and put them in new projects, meaning that making new mashups will only take a fraction of the time.

Let's say for this example that you've got a previous mashup that uses a main track and a breakdown track. The element you want to re-use is the breakdown. You've already done all this work searching for it, importing it, editing it down into its most well-known parts, and you've Warped it to be compatible with dance music tempo. Rather than making a whole new mashup from scratch, the easier approach is to go into your old session and simply swap the old main track for the new one.

You can follow these steps to get started quickly.

Go into your 'old' session and find the old project. Move the folder out of the 'old' folder and back up into the main directory, to remind you that it is an active session again. Open up the session in Live.

As you look at the Arrangement area, you'll be reminded of what you originally used in the mashup. Let's say it looks like the one in Figure 10.1.

The first thing you should do is go to File > Save Live Set As and save it as a new project file. Do this right away so that you don't overwrite your previous mashup. Then take a look at the session. In this example we have a main track, a breakdown track, some effects and 'hinting' loops.

Think about whether you want to keep your effects. Usually they are in there because some part of the breakdown track or transitions needed help being

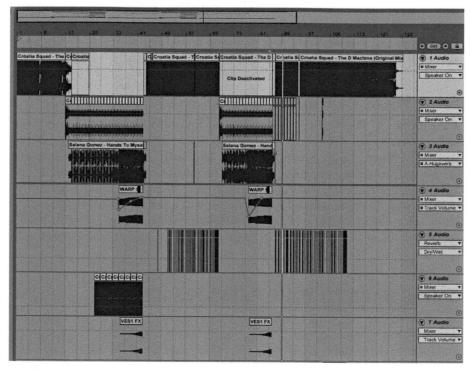

FIGURE 10.1
Returning to a Previous Mashup to Harvest an Element

Screenshot Appears Courtesy of Ableton

energetic and exciting, so while it's usually a good idea to leave them in at first, you should come back later and check whether they're still necessary. In this example we will just mute the main track (most easily done by simply muting the Audio Track rather than the individual clips).

Now we drag an empty Audio Track up from the bottom of the session and place it just above the main track. Do this by grabbing it by the title (for example '11 Audio'). Drop your *new* main track on the empty track, but place the song right at the end of the timeline, after the mashup. This is because we need to listen to it on its own and determine how its arrangement is different from the old main track. It may have longer or shorter breakdowns, drops or transitions. You may have to make the whole mashup a little longer or shorter, possibly removing a breakdown or even repeating the same 'drop' section in your main track to fit the new arrangement.

So go ahead and find the new element's starting point and set up the Warp markers as per usual. Remember that if your new main track is a different BPM, it's better to change the BPM for the whole session to match it, rather than

Warping your main track. If the new track is a semitone up or down from the old one, perform any key changes you need to, preferably to the breakdown elements.

If you need to make extensions or reductions to the space you have allowed for drops, you will need to duplicate or cut entire sections of 8 across the whole mashup to fit your new drop sections in. If so, select a section of eight bars and use *Duplicate Time* or *Delete Time* under the Edit menu.

Figure 10.2 shows a new 'main track' element (on the top-most Audio Track) sliced up into the desired parts and ready to replace the old element.

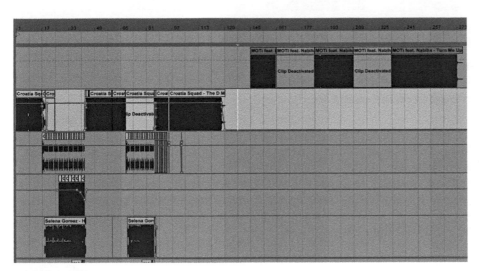

FIGURE 10.2
A New 'Main Track' Standing By

Screenshot Appears Courtesy of Ableton

Once you've decided on your arrangement and made sure the old mashup sections are the right lengths to fit it, create splits (CTRL+E Win, CMD+E Mac) in your new track and move the sections you want into the timeline of the mashup, lining them up with the old main track. Pay special attention to any volume, filter, low-shelf and Send automation used on the old track and be sure to re-make it on the new track. Remember that the new song will be sonically different and will require subtly different automation, so re-create it with the help of your ears rather than simply copying the automation across.

See the example mashup in Figure 10.3, which has a new main track inserted. Small changes needed to be made in the arrangement and the automation to make the new element fit, but most of the hard work was done in the original session.

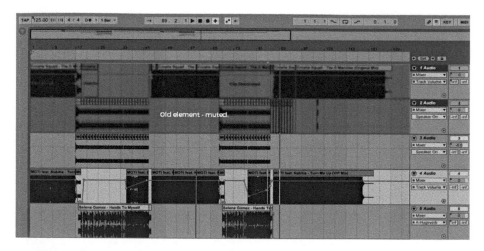

FIGURE 10.3
The Reinvented Mashup with a New 'Main Track'

Screenshot Appears Courtesy of Ableton

This is a very quick way to come up with the foundation for your new mashup. From here, have a careful listen through from start to finish and listen out for anything that has not translated well. Take special notice of bad transitions, too many or too few effects, and important moments that no longer work. Do what you have to do to make your new idea work, and create any new moments in the context of the new main track.

This method is incredibly efficient if you want to re-use more than one element from an old mashup. For example, if you have an old mashup with a main track, a breakdown track, an acapella and a group of effects and builds, but the only thing you want to replace is the main track, think of how much time you are saving yourself by following the above steps as opposed to re-creating the whole mashup.

INTROS AND SET-BREAKERS

Once you have mastered the art of mashup creation, you can use the techniques you have learned to create other useful DJ tools. Think of some interesting ways to kick off your DJ set; such as beginning with a classic track, a huge drum build-up with an acapella over it, attention-grabbing sound effects or mood-setting synth pads.

There are any number of exciting ways you can start things off before you get the audience in the groove of a regular dance tempo, and you don't need to be a full-scale producer to create them. For some DJs, a set intro is about giving themselves time to address the crowd on the microphone without any vocal playing underneath. For others, it is about 'resetting' the dance floor vibe so that they can rebuild it in their own way. For others still, it is about directing attention to the booth as they take over the decks, or creating a memorable beginning to their set. Some begin big festival sets with an almost cinematic introduction, whilst producer/DJs often play a special introduction version of their biggest song.

You already have all the ingredients. You can use drum builds from longer, sparser tracks. You can also add your own drum builds and upsweep effects from your samples folder. You can use modern or classic tracks from your mashup folders. You can use well-known acapellas, vocal chants and loops from your acapella folders. Essentially a set intro is simply a short mashup that has a high building energy and doesn't have an ending.

Think of a few different styles of gig you are likely to play at and come up with a few varied intros to suit them. Some can be short, some can be drawn out. Some may require you to stop the previous DJ's final song by filtering or fading out, others can be mixed into smoothly. The great thing about having multiple set intros to choose from is that you can read the crowd before you go on and begin your set in a way that makes sense. Try to have your intros build for at least one and a half minutes so that you always have time to mix into your first track, no matter what.

Finally, don't be afraid to use the same tool to break up your DJ set. An 'intro-style' track can be a great way to take your set from one energy level to another, or to bring dancers out of their own little world to refocus them on what you are doing on the decks. You can create similar tracks to the above, only with a dedicated intro so that you can mix into them. Then they can drop into effects, slow down or reset the vibe however you want.

CHAPTER 11
Final Thoughts

KEEP YOUR EARS OPEN FOR GREAT IDEAS

In order to advance mashup creation it is essential to check out many ideas from the big artists. Some great examples that I was able to learn many tips and tricks off would be Swedish House Mafia (now defunct, but made some seriously amazing mashups), MAKJ, Henry Fong, Hardwell, Dannic, Arthur White and Mash'd N Kutcher to name a few.

(DJ Trim, Club DJ)

Every time you hear another DJ play mashups, keep your ears open for new concepts. Don't rip off their actual ideas, but let the way they combine or transition between elements be a constant inspiration for new creativity. Listen to DJs from other genres. Each genre has different norms for tempi, rhythm styles or song arrangements, and DJs will possess different techniques to handle them. You can always find ways to give audiences better moments.

DON'T OVERDO IT

Remember—Mashups should NOT be the basis of your ENTIRE set! The effect becomes weakened. Sometimes, people just want to hear the original song.

(James Ash, DJ/Producer)

I think there are a few purposes for using mash ups, however overuse can end up just being a bit tasteless.

(Kam Denny, DJ/Producer)

Often people are caught trying to overcomplicate the process, especially when the tracks you use for mashups are already mastered and purely sound good in most cases without any real need to add any plugins to affect the sounds. Less is always more with mashups.

(DJ Trim, Club DJ)

In the modern dance world, DJs have understandably responded to the growing pressure to use mashups in their sets by trying to make things complicated for the sake of it. Just don't overdo it. It's more and more common to hear DJ sets that sound like a mega-mix. You can see people on the dance floor get excited when they hear something they like, and then disappointed when it transitions out ten seconds later. Try to learn the difference between making a mashup for the benefit of the crowd and making a mashup just for the sake of it. Learn the difference between what's tasteful and what's overdone.

If in doubt, just play the original song.

LEGAL RAMIFICATIONS OF DISTRIBUTING MASHUPS

Mashups are a great way of taking your live sets to the next level. But, always remember that the recordings you are using belong to the original artists and their record labels. Be respectful, and don't distribute your mashups to the public (free or for sale). It's essentially giving away their music, and is illegal in the eyes of the law. Keep them in your personal collection, and play them out as much as you want—then it's a win for everybody. After all, the best thing about creating your own mashups is that they are unique to your sound and style as a DJ. They belong in your set.

MORE RESOURCES

There are a heap of free resources on the 'How To Make Great Music Mashups' website. You can find organisational tools such as mashup checklists templates, an example mashup key list, a selection of drum and effect samples, key guides for every scale and more.

Don't forget that all the videos are there as well. They cover Chapters 6 to 9, from an introduction to Ableton Live through to the techniques I've covered in the book. Just head to 'makegreatmusicmashups.com' to access them.

THANK YOU

Thanks for reading. I hope you have learned a few things that will help you keep your audiences coming back for more.

If you have any questions or feedback, or would even like to share some of your own mashup techniques, I'd be happy to hear from you. Just contact me through the website.

Thanks for being awesome.

Now, get out there and create some great moments.

Resources and Links

Bonus content and follow-along videos: www.makegreatmusicmashups.com

Ableton Live software: www.ableton.com

Ableton Live manual: www.ableton.com/manual/welcome-to-live/

Mixed In Key software: www.mixedinkey.com

Xfer's LFO-Tool plug-in: www.xferrecords.com/products/lfotool

'Limiter No.6' by Vladg Sound: https://vladgsound.wordpress.com/downloads/

Huge collection of acapellas: www.acapellas4u.co.uk

Popular dance music samples for effects, sweeps etc.: www.vengeance-sound.com/samples.php

Dance music's most popular online music distributor: www.beatport.com

Keyboard Shortcuts

Ableton Live Keyboard Shortcuts	Windows	Mac
Views		
Switch between Session and Arrangement view	TAB	TAB
Switch between Clip View and Device View	SHIFT+TAB	SHIFT+TAB
Hide/Show Browser	CTRL+ALT+B	CMD+ALT+B
Hide/Show Sends	CTRL+ALT+S	CMD+ALT+S
Close Window/Dialog	ESC	ESC
Zoom In	+	+
Zoom Out	–	–
Browser		
Load Selected Item from Browser	ENTER	ENTER
Preview Selected File	SHIFT+ENTER	SHIFT+ENTER
Search In Browser	CTRL+F	CMD+F
Transport		
Play (from Start Marker)/Stop	SPACE	SPACE
Move Start Marker along by one unit (depending on zoom level)	LEFT/RIGHT	LEFT/RIGHT
Loop Playback on Selection	CTRL+L	CMD+L
Editing		
Cut	CTRL+X	CMD+X
Copy	CTRL+C	CMD+C
Paste	CTRL+V	CMD+V
Duplicate	CTRL+D	CMD+D
Delete	DELETE	DELETE
Undo	CTRL+Z	CMD+Z
Redo	CTRL+R	CMD+R
Select All	CTRL+A	CMD+A
Bypass Snap while dragging clips	Hold ALT	Hold CMD
Turn Snap To Grid off/on	CTRL+4	CMD+I
Toggle Draw Mode (for automation)	CTRL+B	CMD+B
Create a splice in a clip	CTRL+E	CMD+E
Select multiple clips	Hold SHIFT	Hold SHIFT
Insert Silence	CTRL+I	CMD+4

Keyboard Shortcuts

Ableton Live Keyboard Shortcuts	Windows	Mac
Commands for Tracks		
Create Audio Track	CTRL+T	CMD+T
Create Return Track	CTRL+ALT+T	CMD+ALT+T
Rename Selected Track	CTRL+R	CMD+R
Duplicate Track	CTRL+D	CMD+D
Open context menu in clip/sample view	Right-click	CTRL-Click
Project commands		
New Live Set	CTRL+N	CMD+N
Open Live Set	CTRL+O	CMD+O
Save Live Set	CTRL+S	CMD+S
Save Live Set As . . .	CTRL+SHIFT+S	CMD+SHIFT+S
Export Audio/Video	CTRL+SHIFT+R	CMD+SHIFT+R
Quit Live	CTRL+Q	CMD+Q

Glossary

Acapella—A piece of audio that contains only the vocal part of a song. From the Latin 'A Cappella'.

Amplitude—The strength of a vibrating wave. In sound, the loudness of the sound; in digital audio, the signal level.

Bar—A term meaning the grouping of a number of beats in music, most often four beats. Also known as a measure.

Bass—The lower range of audio frequencies, also referred to as low-end.

Bass-line—Refers to the lowest playing instrument in the music. Can be the main feature in many genres of dance music.

BPM—Stands for Beats-per-Minute. The measurement for the tempo or speed of a piece of music.

Breakdown—The section in a song where the drums drop out or decrease. In dance music, usually intended to provide the 'story' or theme leading up to dancing sections.

Chorus—The part of a song that is repeated and has the same music and lyrics each time; the chorus will usually give the central story of the song.

Clip (Ableton Live)—Live's name for a single piece of audio or MIDI.

Clipping—The action of deforming a waveform during overload. Samples or signals are unable to be written at their intended amplitude and are flattened against the recordable boundaries.

DAW—Digital Audio Workstation. A computer-run application for digital audio manipulation, such as Ableton Live, Avid ProTools and Apple Logic Studio.

Decay—In terms of reverb, how long it takes the tail of audio to reach −60dB of its original volume.

Decibel (dB)—Relative measurement for the volume (loudness) of sound. Measured as a multiple of the logarithm of the ratio between levels.

Delay—An effects processor that applies repetitions of incoming audio at specific intervals, creating a distance effect. Sometimes referred to as echo.

DIY—Short for 'Do It Yourself'. In terms of acapellas, it means somebody has created their own acapella using a combination of phase-cancellation, filtering and editing.

Drop—A recent nickname for the section of dance music that has drums in it, particularly referring to the first instance of it.

Dry—Describes a sound with no effects processing being applied to it.

Envelope—How a sound or audio signal varies in intensity over a time span.

Equalisation (EQ)—The process of adjusting the tonal quality of a sound, by making volume adjustments to certain frequency bands.

Export—To render a production session into a single destination, such as a WAV file.

Fade—A gradual reduction or rise of the level of the audio signal.

Feedback—The amount of delayed signal sent back to the input of a delay line, used in repeat-echo effects.

Filter—An effects processor that removes signals with frequencies above or below a certain point (called the cut-off frequency).

Flat—One of the black notes on a piano keyboard, one semitone down from its relative white note. Not all white notes have a flat version.

Formant—An element in the sound of a voice or instrument that does not change frequency as different pitches are sounded.

Frequency—A measurement for the number of occurrences per second. In audio, it describes the cycles per second of a waveform.

Gain—The amount of increase in audio signal strength, often expressed in decibels (dB). Reductions are expressed as minus amounts.

High-Cut—Only allows the frequencies below the desired frequency to pass through (also known as Low-Pass).

High-Pass—Only allows the frequencies above the desired frequency to pass through (also known as Low-Cut).

Hz—Abbreviation for Hertz, a unit of measurement that represents frequency, or the number of cycles per second.

Intro—Short for Introduction, used to describe the lead in before a dance track reaches its core sections.

kbps—Kilobytes per second. In audio, usually meant to represent the quality and audio fidelity of an .mp3 file.

Key (music)—The tonic note and chord that gives a subjective sense of arrival and rest. The chord in a piece of music that feels like the 'home' or 'centre' of the song.

kHz—An abbreviation of kilo-Hertz, i.e. 5kHz = 5000Hz. Used to make large Hz values more manageable.

Kick Drum—Another term for Bass Drum. In dance music, this usually is the loudest and most important drum element.

Level—The amount of signal strength; the amplitude, especially the average amplitude. Often referred to as volume.

LFO—Low Frequency Oscillator. Its signal can be used as a source to create looping patterns that affect the values of other audio signals such as level and filtering.

Limiter—A severe and fast-reacting compressor designed to protect audio hardware and systems by preventing levels above a certain amplitude.

Loop—A repeating section of audio where each repetition is identical.

Low-Pass—Only allows the frequencies below the desired frequency to pass through (also known as High-Cut).

Low-Cut—Only allows the frequencies above the desired frequency to pass through (also known as High-Pass).

Major Scale—A selection of notes that often make a piece of music feel happy, excited or playful.

MIDI—Short for Musical Instrument Digital Interface; a digital signal system (a system of number signals) used to communicate performance information to and from musical instruments making music.

Minor Scale—A selection of notes that often make a piece of music feel sad, cool or thoughtful.

Mix—The signal made by blending individual signals together.

Monitors—The speakers facing back onto the stage; or in the case of studios, the main playback speakers.

mp3—A means of compressing a sound sequence into a very small file, to enable digital storage and transmission. Much of modern dance music is now stored in .mp3 format.

Mute—To temporarily prevent a track's audio signal from reaching the master channel.

Octave—A difference of pitch where one tone has a frequency that is double or one-half of the frequency of another tone. Interpreted by the ear as the 'same note', at different heights of pitch.

Outro—The final and concluding section of a piece of music.

Peak—The highest point in the audio waveform.

Pitch—The perception of frequency by the ear (a higher or lower property of music).

Preset—A set of factory-set parameters to give one effect on a signal processing device.

Q—The narrowness or broadness of the frequencies affected by an EQ change. Also referred to as resonance.

Remix—An alternate version of a song where a producer (usually not the original artist) has re-worked some or all the available audio stems to present the song in a different arrangement or style.

Resonance—The narrowness or broadness of the frequencies affected by an EQ change. Also referred to as 'Q'.

Reverb—Short for reverberation. The persistence of a sound after the source stops emitting it, caused by many discrete echoes arriving at the ear so closely spaced in time that the ear cannot separate them.

Ride—Short for Ride Cymbal, a high-frequency drum often used in dance music to increase energy without overtly changing the rhythm.

Sampling (audio editing)—To record a short segment of audio for the purpose of playback later.

Sample (digital recording)—To measure the level of a waveform at a given instant.

Scale—In music theory, a scale is any set of musical notes ordered by fundamental frequency or pitch. Usually defined as a set pattern of semitones and tones.

Semitone—The distance between one musical note and its closest neighbouring note.

Sharp—One of the black notes on a piano keyboard, one semitone up from its relative white note. Not all white notes have a sharp version.

Shelf—A frequency response of an equalisation circuit where the boost or cut of frequencies forms a shelf on a frequency response graph.

Side-Chain Compression—A form of compression where the affected audio receives its compression trigger from an alternate audio signal.

Snare—Short for Snare Drum, the medium size drum directly in front of a sitting drummer that has metal strands drawn across the bottom head that rattle when the drum is hit.

Solo—A circuit in a console (or control in a DAW) that allows just one channel (or several selected channels) to be heard or to reach the output with all others muted.

Splice—To create an edit point and split a piece of audio into two, so as to move them independently of each other.

Stereo—A recording or reproduction of at least two channels where positioning of instrument sounds left to right can be perceived.

Sync—The running of two devices or audio streams in time with one another.

Synthesiser (or Synth)—A musical instrument that artificially (using oscillators) generates signals to simulate the sounds of real instruments or to create other sounds not possible with real instruments.

Tempo—The rate at which the music moves measured in Beats-per-Minute (how many steady even pulses there are in the music per minute).

Time Signature—The rhythmical properties of a piece of music. It defines how many beats appear in a bar (or measure) of music, and determines the subdivisions of those beats.

Tone—Twice the distance between one musical note and its closest neighbouring note.

Transient—The initial high-energy peak at the beginning of a waveform, such as one caused by the percussive action of a pick or hammer hitting the string, etc.

Transition—The process of moving from one piece of music to another.

Vocal—A musical performance by the voice, usually singing.

Volume—See 'Level'.

Warp, Warping—The function in Ableton Live that allows one to change the natural speed of audio without changing the pitch, and vice versa.

wav—Waveform Audio File Format, a standard file format for storing an audio bitstream on digital Devices.

Waveform—The shape made by the fluctuations of a quantity over time. The waveform is displayed graphically on audio clips in most DAWs.

Wet—An audio signal that is completely affected by a signal processor and contains none of the original 'dry' signal.

Window—A portion of a file shown on a screen, usually appearing as a menu on top of the current page of data.

Index

Milton Keynes UK
Ingram Content Group UK Ltd.
UKHW051923141024
449569UK00027B/1333